Math Fun Grade 4

S0-DJE-492
Best Value Books™

Table of Contents

The student pages in this book have been specially prepared for reproduction on any standard copying machine.

Kelley Wingate products are available at fine educational supply stores throughout the U.S. and Canada.

ISBN 0-88724-442-4

About the book...

This book is just one in our Best Value™ series of reproducible, skill oriented activity books. Each book is developmentally appropriate and contains over 100 pages packed with educationally sound classroom-tested activities. Each book also contains free skill cards and resource pages filled with extended activity ideas.

The activities in this book have been developed to help students master the basic skills necessary to succeed in mathematics. The activities have been sequenced to help insure successful completion of the assigned tasks, thus building positive self-esteem as well as the self-confidence students need to meet academic and social challenges.

The activities may be used by themselves, as supplemental activities, or as enrichment material for the mathematics program.

Developed by teachers and tested by students, we never lost sight of the fact that if students don't stay motivated and involved, they will never truly grasp the skills being taught on a cognitive level.

About the author...

Dawn Talluto Jacobi holds a Bachelor's degree in Mathematics. While raising her three children (Eric, Kaitlin, and Matthew) Dawn found herself in demand as a math tutor. She noted a common thread that kept children from finding success in math class - the lack of self-confidence. Dawn developed a game format that attracted and held her students' attention because it made math more fun. Dawn discovered that she enjoyed teaching and decided to enter the field full time. She is currently teaching high school Algebra and is working on her Master's degree in Education.

Senior Editors: Patricia Pedigo and Dr.Roger De Santi
Production Director: Homer Desrochers
Production: Arlene Evitts and Debra Ollier

Ready-To-Use Ideas and Activities

The activities in this book will help children master the basic skills necessary to become competent learners. Remember, as you read through the activities listed below and as you go through this book, that all children learn at their own rate. Although repetition is important, it is critical that we never lose sight of the fact that it is equally important to build children's self-esteem and self-confidence if we want them to become successful learners.

Flashcard ideas

The back of this book has removable flashcards that will be great for basic skill and enrichment activities. Pull the flashcards out and cut them apart (if you have access to a paper cutter, use that). Following are several ideas for use of the flashcards.

- Use the flashcards to practice and reinforce multiplication and division facts. Always remember to tell students the number of problems that were answered correctly. For example, if the student answers 8 problems correctly out of ten, tell the student that he or she got 8 problems correct, not "you missed three". Building self-confidence and fostering good feelings about learning is very important.

- Give each child three or four flashcards. Call out number sentences and have students hold up the correct answer card.

- Turn the flashcards with the number sentence showing. Have students match equivalent number sentences. Self-check by looking at the answers on the back of the card.

- Play team "High-Low". Divide the cards into two piles. Two students turn over the top two cards and answer the number sentence. The player with the highest answer takes both cards. Pass the pile on to the next student and repeat. The team with the most cards at the end wins.

Ready-To-Use Ideas and Activities

Multiplication Bee

Try this twist to a classroom favorite. Using the flashcards in the back of this book, stack all of the multiplication cards in a deck with the problem facing up. Line all students up and ask the first student to answer a multiplication problem. If the student answers it correctly, then he or she remains standing. Students that do not answer correctly, return to their desks. The student left standing is the winner.

Multiplication Race

Divide students into three groups. Have each group line up. Using the flash - cards in the back of this book, stack all of the multiplication cards in a deck with the problem facing up. Ask the first person in each group to walk to the chalkboard and write the answer to a problem that you call out, if the answer is correct, that group receives a point. The group with the most points at the end of the game wins.

Reproduce the bingo sheet included in this book, making enough to have one for each student. Hand them out to the students. Take the flashcards and write the problems on the chalk board. Have the students choose 24 of the problems and write them in any order on the empty spaces of their bingo cards, writing only one problem in each space. When all students have finished filling out their bingo cards, take the flash-cards and make them in to a deck. Call out the answers one at a time. Any student who has a problem that equals the called out answer should make an "X" through the problem to cross it out. The student who crosses out five problems in a row first (horizontally, vertically, or diagonally) wins the game and shouts "BINGO!". Another fun version of this game is to write answers on the bingo cards and call out the problems. To extend the game you can continue playing until you a student crosses out all of the problems on his bingo sheet.

Challenge your own score! The next two pages include basic multiplication and division problems which children should memorize. To help them we suggest you make multiple copies of these pages. Work on only one page at a time. Get a minute timer. See how many problems the child can do correctly in one minute. Record the child's score on a piece of paper. Let the child try again and see how many problems he/she can do correctly. The more times a child does each page, the higher his/her score will become and the more problems he/she will learn. As scores increase so does a child's self-confidence.

1. 3⟌27

2. 2⟌6

3. 5⟌15

4. 3⟌12

5. 8⟌24

6. 9⟌45

7. 8⟌48

8. 4⟌16

9. 2⟌10

10. 7⟌42

11. 9⟌72

12. 4⟌28

13. 2⟌14

14. 5⟌25

15. 6⟌42

16. 7⟌63

17. 8⟌64

18. 5⟌20

19. 2⟌8

20. 5⟌10

21. 8⟌40

22. 9⟌36

23. 2⟌12

24. 3⟌21

25. 4⟌32

26. 7⟌35

27. 6⟌30

28. 4⟌20

29. 9⟌54

30. 7⟌28

31. 9⟌18

32. 6⟌18

33. 4⟌8

34. 7⟌49

35. 5⟌35

36. 3⟌24

37. 2⟌18

38. 5⟌45

39. 9⟌81

40. 5⟌15

41. 7⟌14

42. 4⟌12

43. 3⟌9

44. 4⟌36

45. 6⟌36

46. 8⟌56

47. 6⟌12

48. 9⟌63

49. 8⟌72

50. 5⟌40

51. 9⟌27

52. 6⟌48

53. 8⟌16

54. 4⟌24

55. 3⟌6

56. 8⟌32

57. 6⟌54

58. 7⟌56

59. 5⟌30

60. 2⟌16

61. 3⟌18

62. 6⟌24

63. 7⟌21

64. 2⟌4

65. 6⟌30

66. 8⟌16

67. 2⟌10

68. 3⟌24

69. 9⟌45

70. 4⟌12

71. 7⟌49

72. 5⟌45

iii

CD-3724

1. 1 x 1	2. 3 x 9	3. 4 x 1	4. 7 x 2	5. 1 x 3	6. 4 x 2	7. 7 x 3	8. 4 x 5
9. 6 x 8	10. 2 x 0	11. 1 x 2	12. 1 x 7	13. 8 x 4	14. 7 x 5	15. 2 x 3	16. 4 x 4
17. 7 x 1	18. 1 x 8	19. 2 x 2	20. 4 x 3	21. 2 x 6	22. 6 x 7	23. 1 x 4	24. 7 x 4
25. 6 x 9	26. 1 x 6	27. 7 x 6	28. 9 x 0	29. 2 x 1	30. 1 x 5	31. 6 x 6	32. 3 x 8
33. 5 x 6	34. 1 x 0	35. 2 x 8	36. 3 x 4	37. 4 x 0	38. 1 x 9	39. 6 x 4	40. 7 x 7
41. 6 x 1	42. 2 x 5	43. 7 x 8	44. 5 x 5	45. 2 x 4	46. 3 x 6	47. 5 x 0	48. 5 x 4
49. 3 x 3	50. 4 x 8	51. 7 x 9	52. 4 x 9	53. 5 x 8	54. 5 x 3	55. 6 x 5	56. 2 x 7
57. 5 x 2	58. 5 x 7	59. 2 x 9	60. 9 x 1	61. 6 x 3	62. 4 x 7	63. 7 x 0	64. 5 x 9
65. 6 x 2	66. 5 x 1	67. 3 x 7	68. 8 x 3	69. 4 x 6	70. 6 x 3	71. 8 x 1	72. 8 x 2

CD-3724

Reproduce this page and make your own math bingo game! Use in conjunction with the enclosed flash cards. Popular formats: caller calls equation and students mark answers or caller calls answers and students mark correct equations.

BINGO

		Free!		

v

CD-3724

Hundreds Chart

1	2	3	4	5	6	7	8	9	10
11	12	13	14	15	16	17	18	19	20
21	22	23	24	25	26	27	28	29	30
31	32	33	34	35	36	37	38	39	40
41	42	43	44	45	46	47	48	49	50
51	52	53	54	55	56	57	58	59	60
61	62	63	64	65	66	67	68	69	70
71	72	73	74	75	76	77	78	79	80
81	82	83	84	85	86	87	88	89	90
91	92	93	94	95	96	97	98	99	100

Multiplication and Division Table

x / ÷	1	2	3	4	5	6	7	8	9
1	1	2	3	4	5	6	7	8	9
2	2	4	6	8	10	12	14	16	18
3	3	6	9	12	15	18	21	24	27
4	4	8	12	16	20	24	28	32	36
5	5	10	15	20	25	30	35	40	45
6	6	12	18	24	30	36	42	48	54
7	7	14	21	28	35	42	49	56	63
8	8	16	24	32	40	48	56	64	72
9	9	18	27	36	45	54	63	72	81

RAP

Add the following problems. Do not forget to regroup when necessary.

68	56	83	77	48
+ 34	+ 97	+ 58	+ 84	+ 92

184	556	964	829	726
+ 717	+ 385	+ 328	+ 347	+ 478

851	365	982	735	578
+ 579	+ 498	+ 469	+ 695	+ 632

3,842	5,096	8,147	7,865	8,907
+ 6,787	+ 2,988	+ 1,676	+ 4,348	+ 4,537

6,469 + 4,153 =   3,967 + 8,946 =

8,355 + 4,987 =   3,234 + 7,966 =

5,463 + 7,549 =   6,516 + 6,488 =

6,308 + 8,942 =   3,377 + 9,642 =

3,845 + 8,628 =   4,856 + 5,865 =

       CD-3724

RAP
Add the following problems. Do not forget to regroup when necessary.

37	74	69	55	88
48	65	78	87	37
+ 92	+ 38	+ 36	+ 32	+ 86

127	146	847	388	982
192	523	453	947	637
+ 253	+ 978	+ 796	+ 365	+ 256

576	822	157	521	712
834	739	495	709	516
+ 429	+ 245	+ 633	+ 483	+ 294

469 + 153 + 333 =   967 + 946 + 254 =

355 + 987 + 613 =   234 + 966 + 863 =

463 + 549 + 243 =   516 + 488 + 119 =

308 + 942 + 876 =   377 + 642 + 629 =

845 + 628 + 458 =   856 + 865 + 138 =

RAP

Add the following problems. Do not forget to regroup when necessary.

45	47	96	78	58
29	56	85	54	73
+ 57	+ 83	+ 69	+ 73	+ 69

661	465	478	823	829
219	352	509	756	763
+ 452	+ 879	+ 382	+ 417	+ 127

589	255	387	857	229
473	419	876	982	802
+ 831	+ 799	+ 624	+ 156	+ 560

648 + 355 + 771 =   697 + 164 + 543 =

535 + 675 + 378 =   432 + 699 + 368 =

346 + 509 + 483 =   558 + 833 + 257 =

398 + 159 + 725 =   733 + 412 + 604 =

548 + 678 + 248 =   635 + 298 + 773 =

   CD-3724

RSP

Subtract the following problems. Do not forget to regroup when necessary.

457	809	515	300	921
- 186	- 372	- 257	- 197	- 569

324	698	914	736	815
- 187	- 389	- 258	- 497	- 428

802	774	512	439	705
- 339	- 269	- 467	- 358	- 337

603	881	840	637	724
- 498	- 512	- 648	- 588	- 436

929 - 464 =   952 - 626 =

808 - 749 =   723 - 674 =

630 - 252 =   564 - 375 =

700 - 411 =   444 - 228 =

830 - 541 =   846 - 598 =

RSP

Subtract the following problems. Do not forget to regroup when necessary.

1,868 - 799	2,452 - 1,875	8,025 - 3,562	7,524 - 4,965	9,342 - 4,786
9,617 - 4,828	3,806 - 1,497	7,320 - 4,566	9,064 - 2,278	5,436 - 4,557
8,914 - 4,937	5,088 - 1,329	7,307 - 6,798	4,582 - 2,776	9,520 - 3,745
4,562 - 1,856	6,407 - 3,819	5,200 - 4,982	8,315 - 5,276	9,028 - 4,346

$7{,}072 - 6{,}884 =$ $8{,}004 - 5{,}645 =$

$9{,}426 - 2{,}558 =$ $5{,}735 - 1{,}768 =$

$6{,}421 - 3{,}757 =$ $7{,}015 - 2{,}348 =$

$6{,}624 - 1{,}999 =$ $4{,}126 - 1{,}689 =$

$7{,}327 - 5{,}698 =$ $5{,}321 - 3{,}993 =$

RSP

Subtract the following problems. Do not forget to regroup when necessary.

3,304	5,100	6,321	7,034	8,723
- 1,178	- 2,487	- 4,375	- 3,158	- 2,876

9,014	5,601	3,040	8,007	5,432
- 3,658	- 4,996	- 2,183	- 5,389	- 2,848

36,050	32,604	39,701	53,600	81,335
- 18,765	- 15,816	- 15,082	- 38,714	- 33,567

20,306	80,040	73,148	90,004	41,025
- 17,428	- 69,367	- 42,349	- 73,665	- 28,357

5,134 - 1,458 =   7,550 - 2,676 =

8,044 - 3,769 =   8,266 - 6,874 =

20,603 - 16,799 =   53,119 - 13,382 =

51,267 - 28,478 =   38,411 - 19,904 =

27,185 - 11,786 =   46,323 - 18,775 =

       CD-3724

RSP

Subtract the following problems. Do not forget to regroup when necessary.

3,080	7,002	5,401	8,984	5,004
- 1,982	- 3,694	- 4,378	- 6,896	- 2,786

18,463	47,031	70,306	51,006	90,340
- 9,675	- 39,554	- 42,957	- 38,498	- 68,565

80,001	40,102	73,004	40,300	80,404
- 78,327	- 23,487	- 59,775	- 29,472	- 55,652

320,432	263,143	631,227	554,654	723,072
- 179,385	- 177,418	- 580,669	- 165,988	- 395,184

$7,432 - 5,554 =$ $8,436 - 2,958 =$

$48,702 - 13,825 =$ $84,251 - 26,744 =$

$70,345 - 42,557 =$ $93,126 - 57,849 =$

$153,424 - 133,667 =$ $325,471 - 199,684 =$

$237,143 - 181,678 =$ $416,314 - 278,745 =$

RAP/RSP

Add or subtract the following problems. Do not forget to regroup when necessary.

701	684	578	900	452
- 482	+ 398	+ 922	- 187	- 378

5,015	4,402	5,917	8,003	5,627
+ 3,276	- 1,986	+ 5,687	- 2,576	+ 9,359

4,321	5,917	7,403	5,715	9,008
+ 9,821	+ 2,393	- 6,776	+ 3,892	- 4,329

7,092	3,072	6,336	4,624	9,723
- 2,296	+ 7,984	- 3,597	+ 8,835	- 4,738

8,256 - 2,567 =   5,884 + 3,726 =

4,722 + 9,306 =   9,313 - 1,838 =

7,112 - 6,207 =   2,659 + 7,482 =

6,345 - 2,857 =   6,258 + 8,733 =

5,169 + 6,828 =   4,027 - 1,898 =

CD-3724

RAP/RSP

Add or subtract the following problems. Do not forget to regroup when necessary.

702	327	840	761	497
- 554	+ 492	- 338	- 286	+ 615

4,749	9,006	7,668	8,254	7,449
+ 8,298	- 3,577	+ 3,942	- 6,268	+ 7,258

30,408	43,576	58,000	35,542	50,102
- 19,219	+ 87,327	- 42,315	+ 18,991	- 36,826

900,645	987,645	865,098	540,723	319,459
- 382,569	- 398,936	- 258,219	+ 369,288	+ 274,553

723 - 557 =   38,965 + 72,558 =

842 + 396 =   39,457 + 51,338 =

5,332 + 1,768 =   72,436 - 27,597 =

8,104 - 5,448 =   234,503 + 118,262 =

81,406 - 59,818 =   642,303 - 154,614 =

9   CD-3724

RMP

How quickly can you complete this page? Time yourself. Ready, set, go!

Time : _____

Number Correct : _____

34	67	42	58	29	74	33	68
x 6	x 3	x 4	x 5	x 4	x 2	x 5	x 3

83	49	57	39	58	98	67	45
x 6	x 2	x 5	x 3	x 4	x 5	x 6	x 3

27	43	29	73	58	63	92	46
x 6	x 5	x 4	x 5	x 4	x 6	x 5	x 5

55	38	27	42	57	83	78	56
x 4	x 6	x 5	x 6	x 6	x 5	x 6	x 4

19 x 4 =   48 x 5 =   27 x 3 =

67 x 6 =   53 x 3 =   54 x 4 =

71 x 5 =   63 x 6 =   45 x 6 =

37 x 5 =   82 x 4 =   29 x 3 =

68 x 4 =   43 x 3 =   69 x 6 =

Name _____

RMP

How quickly can you complete this page? Time yourself. Ready, set, go!

Time : _____

Number Correct : _____

69	48	57	36	73	95	89	27
x 5	x 4	x 6	x 7	x 3	x 4	x 3	x 8

52	33	82	95	59	44	48	77
x 5	x 6	x 4	x 7	x 3	x 6	x 5	x 4

94	76	47	79	27	88	67	85
x 6	x 3	x 5	x 4	x 6	x 7	x 5	x 4

65	43	29	50	71	38	84	77
x 2	x 6	x 7	x 5	x 4	x 3	x 5	x 6

49 x 4 =   27 x 5 =   71 x 6 =

39 x 3 =   45 x 7 =   54 x 2 =

65 x 4 =   71 x 3 =   39 x 5 =

83 x 4 =   48 x 6 =   63 x 7 =

24 x 4 =   65 x 3 =   59 x 5 =

11   CD-3724

Skill: Regrouping Multiplication

RMP

How quickly can you complete this page? Time yourself. Ready, set, go!

Time : _____

Number Correct : _____

47	53	16	28	58	62	37	24
x 8	x 9	x 8	x 9	x 6	x 7	x 7	x 8

28	37	54	82	43	29	62	91
x 9	x 6	x 5	x 8	x 9	x 7	x 4	x 6

43	48	29	37	49	51	45	37
x 9	x 7	x 6	x 5	x 4	x 7	x 9	x 6

72	46	64	63	28	82	51	23
x 8	x 7	x 4	x 7	x 6	x 8	x 5	x 7

71 x 4 =   37 x 9 =   63 x 7 =

45 x 6 =   66 x 5 =   84 x 6 =

77 x 8 =   59 x 7 =   91 x 6 =

56 x 4 =   38 x 4 =   48 x 5 =

58 x 7 =   78 x 9 =   62 x 9 =

             CD-3724

Name _____

RMP

How quickly can you complete this page? Time yourself. Ready, set, go!

Time : _____
Number Correct : _____

67	92	73	48	92	63	58	73
x 9	x 7	x 8	x 7	x 9	x 6	x 8	x 7

53	48	78	89	53	48	55	87
x 9	x 6	x 5	x 8	x 9	x 7	x 5	x 6

28	56	83	72	56	81	53	88
x 9	x 7	x 6	x 5	x 4	x 7	x 9	x 8

42	35	44	77	94	28	39	99
x 8	x 7	x 4	x 7	x 6	x 8	x 5	x 9

71 x 9 =   25 x 8 =   52 x 6 =

34 x 5=   75 x 4 =   93 x 5 =

68 x 7 =   68 x 6 =   82 x 5 =

65 x 9 =   47 x 9 =   39 x 4 =

69 x 6 =   67 x 8 =   53 x 8 =

Multiply! Multiply!

Multiply the following problems. Do not forget to regroup when necessary.

| 40 | 59 | 84 | 76 | 90 |
| x 38 | x 15 | x 35 | x 38 | x 67 |

| 24 | 96 | 72 | 85 | 21 |
| x 86 | x 33 | x 64 | x 17 | x 39 |

| 48 | 36 | 54 | 32 | 53 |
| x 27 | x 72 | x 56 | x 49 | x 47 |

| 80 | 36 | 29 | 32 | 53 |
| x 98 | x 52 | x 16 | x 44 | x 45 |

69 x 41 =   37 x 46 =

55 x 17 =   34 x 26 =

43 x 49 =   16 x 88 =

58 x 22 =   37 x 42 =

45 x 28 =   56 x 25 =

C

Name _____

Multiply! Multiply!

Multiply the following problems. Do not forget to regroup when necessary.

37 x 12	15 x 20	43 x 25	61 x 38	44 x 94
75 x 26	29 x 94	37 x 60	53 x 84	70 x 16
19 x 67	98 x 32	54 x 53	74 x 59	67 x 38
62 x 16	54 x 30	68 x 22	83 x 32	71 x 35

96 x 12 =   48 x 25 =

64 x 26 =   45 x 15 =

54 x 38 =   27 x 72 =

47 x 31 =   48 x 33 =

56 x 17 =   64 x 36 =

996 Kelley Wingate Publications   CD-3724

Multiply! Multiply!

Multiply the following problems. Do not forget to regroup when necessary.

405	315	923	337	560
x 32	x 74	x 85	x 56	x 62

617	479	982	189	303
x 75	x 24	x 37	x 45	x 41

534	457	816	374	786
x 83	x 94	x 56	x 13	x 48

295	315	447	379	386
x 64	x 32	x 25	x 14	x 42

179 x 43 =   317 x 52 =

825 x 19 =   364 x 27 =

452 x 28 =   426 x 42 =

638 x 36 =   328 x 49 =

145 x 44 =   276 x 32 =

Multiply! Multiply!

Multiply the following problems. Do not forget to regroup when necessary.

615	658	396	447	827
x 37	x 26	x 42	x 67	x 43

513	365	412	307	724
x 74	x 29	x 48	x 75	x 59

382	391	188	507	129
x 86	x 42	x 92	x 68	x 38

767	652	724	883	732
x 35	x 43	x 53	x 64	x 79

459 x 37 =   228 x 32 =

946 x 58 =   673 x 28 =

179 x 36 =   875 x 49 =

286 x 54 =   378 x 58 =

546 x 94 =   638 x 78 =

Name _____

Dandy Division
Divide the following problems. Show your work.

$5\overline{)25}$ $7\overline{)28}$ $9\overline{)54}$ $6\overline{)48}$ $4\overline{)36}$

$8\overline{)40}$ $5\overline{)55}$ $8\overline{)64}$ $7\overline{)49}$ $9\overline{)18}$

$5\overline{)30}$ $12\overline{)60}$ $11\overline{)77}$ $4\overline{)48}$ $10\overline{)90}$

$6\overline{)60}$ $8\overline{)40}$ $4\overline{)36}$ $6\overline{)42}$ $3\overline{)36}$

$63 \div 7 =$ $15 \div 3 =$ $55 \div 5 =$

$45 \div 9 =$ $48 \div 8 =$ $16 \div 4 =$

$12 \div 2 =$ $35 \div 7 =$ $56 \div 8 =$

$32 \div 4 =$ $99 \div 9 =$ $28 \div 7 =$

$20 \div 5 =$ $42 \div 6 =$ $18 \div 3 =$

Dandy Division

Divide the following problems. Show your work.

$4\overline{)44}$ $9\overline{)99}$ $2\overline{)74}$ $6\overline{)90}$ $3\overline{)45}$

$7\overline{)70}$ $5\overline{)80}$ $3\overline{)57}$ $4\overline{)60}$ $9\overline{)90}$

$6\overline{)78}$ $8\overline{)80}$ $7\overline{)91}$ $6\overline{)60}$ $4\overline{)72}$

$3\overline{)39}$ $5\overline{)55}$ $4\overline{)76}$ $8\overline{)72}$ $9\overline{)63}$

$42 \div 2 =$ $75 \div 3 =$ $56 \div 4 =$

$65 \div 5 =$ $72 \div 6 =$ $84 \div 7 =$

$88 \div 8 =$ $90 \div 9 =$ $96 \div 8 =$

$77 \div 7 =$ $84 \div 6 =$ $75 \div 5 =$

$48 \div 4 =$ $54 \div 3 =$ $54 \div 2 =$

Dandy Division

Divide the following problems. Show your work.

3 | 333   5 | 305   8 | 856   7 | 497   9 | 549

5 | 455   8 | 728   6 | 648   3 | 279   2 | 216

7 | 728   9 | 189   4 | 328   3 | 912   2 | 642

459 ÷ 9 =	219 ÷ 3 =
864 ÷ 8 =	628 ÷ 2 =
126 ÷ 6 =	714 ÷ 7 =
288 ÷ 4 =	168 ÷ 8 =
545 ÷ 5 =	426 ÷ 6 =

Dandy Division

Divide the following problems. Show your work.

$6\overline{)726}$ $8\overline{)256}$ $8\overline{)168}$ $8\overline{)968}$ $9\overline{)558}$

$7\overline{)854}$ $5\overline{)225}$ $7\overline{)364}$ $9\overline{)387}$ $3\overline{)552}$

$5\overline{)735}$ $4\overline{)576}$ $6\overline{)384}$ $3\overline{)852}$ $7\overline{)301}$

$336 \div 8 =$ $224 \div 7 =$

$468 \div 9 =$ $628 \div 4 =$

$180 \div 4 =$ $252 \div 6 =$

$228 \div 3 =$ $192 \div 8 =$

$735 \div 5 =$ $258 \div 3 =$

Name _____

Dandy Division

Divide the following problems. Show your work.

$3\overline{)28}$ $3\overline{)14}$ $6\overline{)51}$ $4\overline{)31}$ $2\overline{)17}$

$7\overline{)45}$ $4\overline{)25}$ $2\overline{)11}$ $8\overline{)47}$ $3\overline{)16}$

$6\overline{)20}$ $4\overline{)19}$ $5\overline{)33}$ $3\overline{)26}$ $7\overline{)50}$

$7\overline{)45}$ $2\overline{)17}$ $5\overline{)59}$ $7\overline{)60}$ $6\overline{)33}$

$23 \div 2 =$ $10 \div 3 =$ $21 \div 4 =$

$48 \div 5 =$ $41 \div 6 =$ $37 \div 7 =$

$39 \div 8 =$ $50 \div 9 =$ $52 \div 8 =$

$45 \div 7 =$ $40 \div 6 =$ $33 \div 5 =$

$27 \div 4 =$ $14 \div 3 =$ $11 \div 2 =$

Name _____

Dandy Division

Divide the following problems. Show your work.

9$\overline{)25}$ 6$\overline{)39}$ 7$\overline{)51}$ 9$\overline{)61}$ 2$\overline{)45}$

4$\overline{)54}$ 5$\overline{)64}$ 8$\overline{)89}$ 7$\overline{)93}$ 3$\overline{)29}$

9$\overline{)58}$ 8$\overline{)93}$ 5$\overline{)73}$ 4$\overline{)67}$ 7$\overline{)36}$

7$\overline{)85}$ 9$\overline{)80}$ 3$\overline{)64}$ 9$\overline{)84}$ 6$\overline{)82}$

$63 \div 2 =$ $98 \div 9 =$ $81 \div 7 =$

$53 \div 4 =$ $57 \div 2 =$ $73 \div 4 =$

$96 \div 7 =$ $89 \div 5 =$ $65 \div 8 =$

$67 \div 3 =$ $59 \div 8 =$ $75 \div 9 =$

$93 \div 6 =$ $76 \div 6 =$ $93 \div 2 =$

               CD-3724

Dandy Division

Divide the following problems. Show your work.

$8\overline{)826}$ $7\overline{)449}$ $5\overline{)107}$ $4\overline{)249}$ $3\overline{)334}$

$6\overline{)482}$ $5\overline{)456}$ $8\overline{)849}$ $7\overline{)288}$ $4\overline{)166}$

$3\overline{)320}$ $9\overline{)722}$ $8\overline{)809}$ $9\overline{)458}$ $3\overline{)121}$

$982 \div 9 =$ $629 \div 6 =$

$847 \div 8 =$ $727 \div 7 =$

$253 \div 3 =$ $371 \div 4 =$

$733 \div 6 =$ $298 \div 7 =$

$932 \div 8 =$ $734 \div 9 =$

Name _____

Dandy Division

Divide the following problems. Show your work.

5 ⟌516   4 ⟌327   3 ⟌217   8 ⟌747   6 ⟌629

6 ⟌369   3 ⟌316   6 ⟌374   9 ⟌439   7 ⟌436

9 ⟌921   4 ⟌519   9 ⟌883   4 ⟌275   8 ⟌503

772 ÷ 3 =   297 ÷ 4 =

822 ÷ 5 =   729 ÷ 6 =

376 ÷ 7 =   553 ÷ 8 =

442 ÷ 9 =   351 ÷ 2 =

694 ÷ 7 =   862 ÷ 5 =

       CD-3724

Hopscotch

Use your math facts to complete the hopscotch board.

1.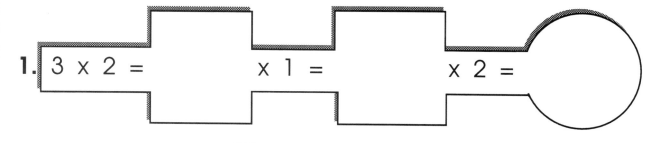
3 x 2 =   x 1 =   x 2 =

2.
2 x 1 =   x 2 =   x 1 =

3.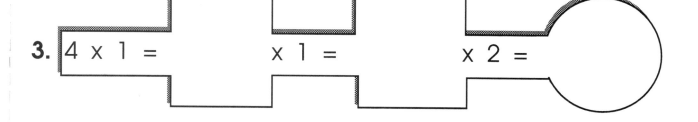
4 x 1 =   x 1 =   x 2 =

4.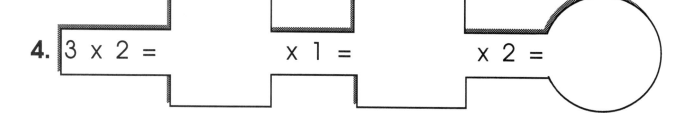
3 x 2 =   x 1 =   x 2 =

5.
2 x 1 =   x 3 =   x 1 =

6.
3 x 2 =   x 3 =   x 1 =

Hopscotch

Use your math facts to complete the hopscotch board.

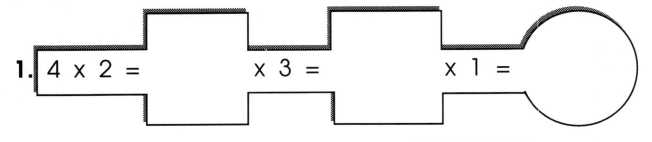

1. 4 x 2 =   x 3 =   x 1 =

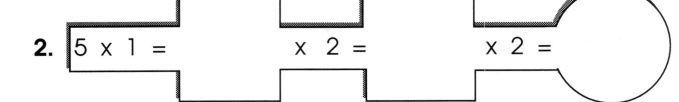

2. 5 x 1 =   x 2 =   x 2 =

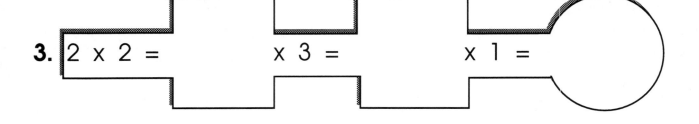

3. 2 x 2 =   x 3 =   x 1 =

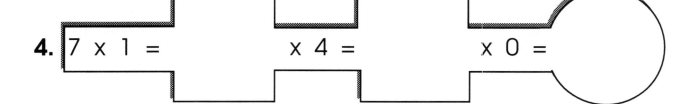

4. 7 x 1 =   x 4 =   x 0 =

5. 4 x 2 =   x 3 =   x 1 =

6. 2 x 4 =   x 2 =   x 2 =

Hopscotch

Use your math facts to complete the hopscotch board.

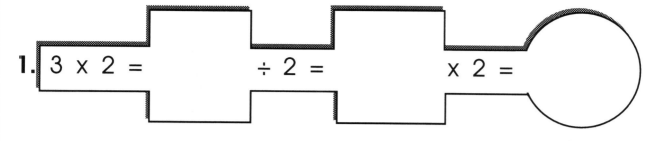

1. $3 \times 2 =$ $\div 2 =$ $\times 2 =$

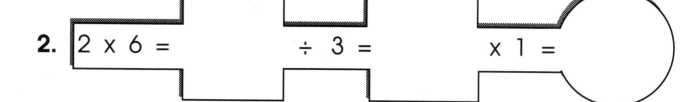

2. $2 \times 6 =$ $\div 3 =$ $\times 1 =$

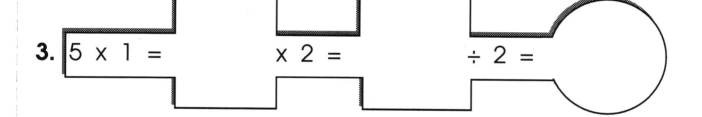

3. $5 \times 1 =$ $\times 2 =$ $\div 2 =$

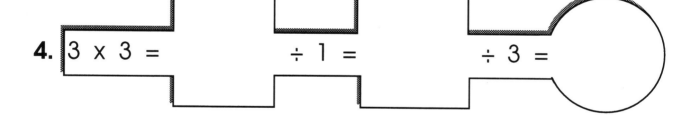

4. $3 \times 3 =$ $\div 1 =$ $\div 3 =$

5. $2 \times 5 =$ $\times 2 =$ $\times 1 =$

6. $3 \times 3 =$ $\times 1 =$ $\times 2 =$

       CD-3724

Name _____

Hopscotch

Use your math facts to complete the hopscotch board.

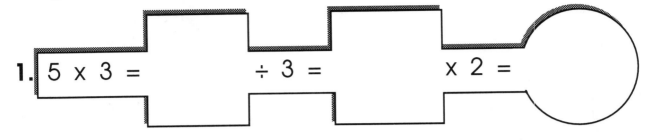

1. $5 \times 3 =$ $\div 3 =$ $\times 2 =$

2. $2 \times 3 =$ $\div 3 =$ $\times 4 =$

3. $6 \times 5 =$ $\times 2 =$ $\div 6 =$

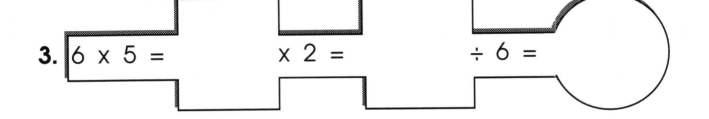

4. $9 \times 3 =$ $\div 3 =$ $\div 9 =$

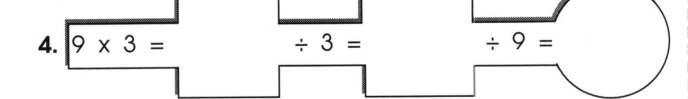

5. $4 \times 5 =$ $\times 3 =$ $\div 10 =$

6. $5 \times 3 =$ $\times 1 =$ $\times 3 =$

CD-3724

Hopscotch

Use your math facts to complete the hopscotch board.

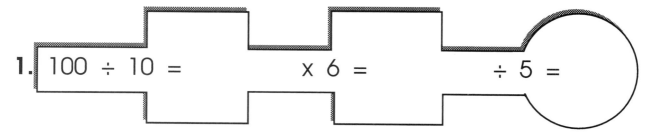

1. | 100 ÷ 10 =   x 6 =   ÷ 5 =

2. | 60 ÷ 5 =   x 4 =   ÷ 8 =

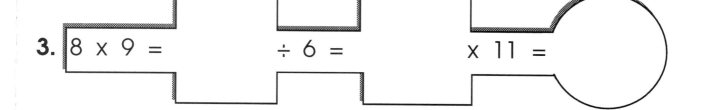

3. | 8 x 9 =   ÷ 6 =   x 11 =

4. | 110 ÷ 11 =   x 3 =   ÷ 6 =

5. | 81 ÷ 9 =   x 4 =   ÷ 12 =

6. | 120 ÷ 10 =   ÷ 2 =   x 9 =

Hopscotch

Use your math facts to complete the hopscotch board.

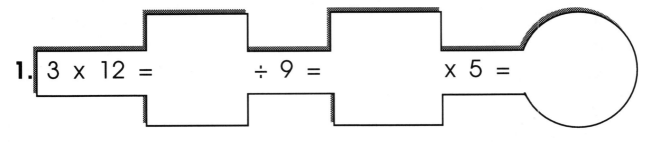

1. 3 x 12 =   ÷ 9 =   x 5 =

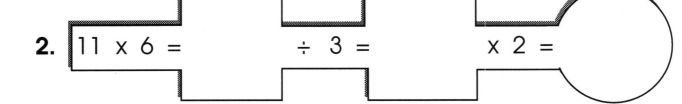

2. 11 x 6 =   ÷ 3 =   x 2 =

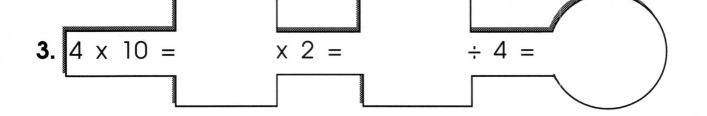

3. 4 x 10 =   x 2 =   ÷ 4 =

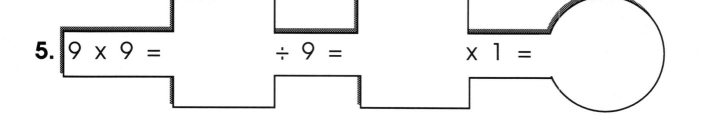

4. 8 x 4 =   ÷ 16 =   x 3 =

5. 9 x 9 =   ÷ 9 =   x 1 =

6. 12 x 3 =   x 2 =   ÷ 8 =

31   CD-3724

Hopscotch

Use your math facts to complete the hopscotch board.

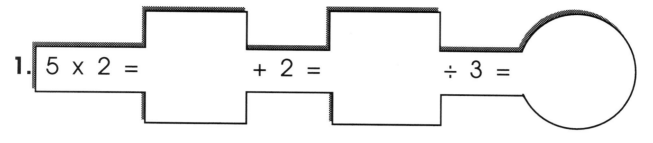

1. 5 x 2 =   + 2 =   ÷ 3 =

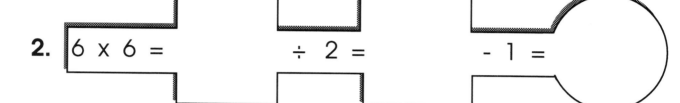

2. 6 x 6 =   ÷ 2 =   − 1 =

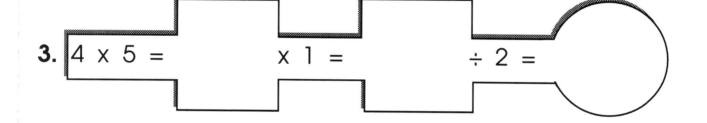

3. 4 x 5 =   x 1 =   ÷ 2 =

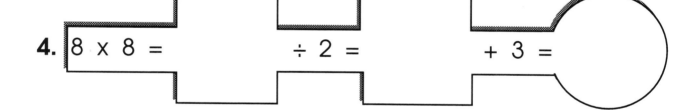

4. 8 x 8 =   ÷ 2 =   + 3 =

5. 27 ÷ 3 =   x 2 =   ÷ 2 =

6. 6 x 8 =   ÷ 12 =   x 5 =

CD-3724

Hopscotch

Use your math facts to complete the hopscotch board.

1. $5 \times 8 =$ $\div 4 =$ $+ 2 =$

2. $9 \times 8 =$ $\div 8 =$ $- 1 =$

3. $12 \div 2 =$ $\times 3 =$ $- 7 =$

4. $3 \times 8 =$ $\div 1 =$ $\div 6 =$

5. $7 \times 7 =$ $+ 1 =$ $\div 2 =$

6. $8 \times 8 =$ $+ 6 =$ $\div 10 =$

Missing Factors

Fill in the box with a number that will make the statement true.

$\boxed{} \div 4 = 7$ $\boxed{} \div 9 = 7$

$\boxed{} \div 8 = 7$ $\boxed{} \div 3 = 6$

$\boxed{} \div 9 = 8$ $\boxed{} \div 7 = 12$

$\boxed{} \div 9 = 9$ $\boxed{} \div 12 = 12$

$\boxed{} \div 12 = 8$ $\boxed{} \div 2 = 12$

$\boxed{} \div 11 = 12$ $\boxed{} \div 8 = 2$

$\boxed{} \div 12 = 5$ $64 \div \boxed{} = 8$

$36 \div \boxed{} = 3$ $60 \div \boxed{} = 12$

$45 \div \boxed{} = 9$ $27 \div \boxed{} = 3$

$54 \div \boxed{} = 9$ $120 \div \boxed{} = 10$

$42 \div 6 = \boxed{}$ $110 \div \boxed{} = 11$

$36 \div 6 = \boxed{}$ $108 \div \boxed{} = 9$

Missing Factors

Fill in the box with a number that will make the statement true.

$\boxed{} \div 10 = 9$

$\boxed{} \div 2 = 7$

$\boxed{} \div 7 = 11$

$\boxed{} \div 4 = 7$

$\boxed{} \div 6 = 6$

$\boxed{} \div 10 = 12$

$36 \div \boxed{} = 9$

$35 \div \boxed{} = 7$

$21 \div \boxed{} = 3$

$18 \div \boxed{} = 6$

$99 \div \boxed{} = 9$

$132 \div \boxed{} = 12$

$\boxed{} \div 6 = 9$

$\boxed{} \div 6 = 12$

$\boxed{} \div 4 = 4$

$\boxed{} \div 9 = 12$

$48 \div \boxed{} = 6$

$56 \div \boxed{} = 8$

$84 \div \boxed{} = 7$

$63 \div \boxed{} = 9$

$49 \div \boxed{} = 7$

$32 \div \boxed{} = 8$

$121 \div \boxed{} = 11$

$144 \div \boxed{} = 12$

Blankety- Blanks

Solve the problems below and write the answer in the box. On the blanket, shade in all the numbers that are in the answer boxes. The answers will make a pattern.

$96 \div 8 \times 5 =$ ☐

$13 - 4 + 6 =$ ☐

$27 \div 3 + 5 =$ ☐

$4 + 7 - 6 =$ ☐

$3 \times 8 \div 12 =$ ☐

$4 \times 9 \div 12 =$ ☐

$40 \div 8 + 6 =$ ☐

$56 \div 7 + 5 =$ ☐

$36 \div 6 \times 9 =$ ☐

$16 - 9 + 5 =$ ☐

$28 \div 4 \times 7 =$ ☐

$24 \div 2 \div 2 =$ ☐

$72 \div 8 \times 2 =$ ☐

$2 \times 7 - 5 =$ ☐

$120 \div 10 - 4 =$ ☐

$15 \div 3 \times 5 =$ ☐

2	3	1	4
10	5	6	7
16	17	8	9
19	24	12	11
20	13	14	21
18	15	23	22
25	26	27	28
49	30	32	33
54	40	45	47
60	55	57	61

Name_____

Skill: All Operations

Blankety- Blanks

Solve the problems below and write the answer in the box. On the blanket, shade in all the numbers that are in the answer boxes. The answers will make a pattern.

4 x 5 ÷ 10 = ☐

32 ÷ 8 x 4 = ☐

2 x 4 x 8 = ☐

5 x 8 ÷ 4 = ☐

77 ÷ 11 + 6 = ☐

96 ÷ 8 - 7 = ☐

8 x 3 ÷ 4 = ☐

63 ÷ 9 + 8 = ☐

55 ÷ 5 - 4 = ☐

144 ÷ 12 x 9 = ☐

14 - 6 + 3 = ☐

2 x 3 x 8 = ☐

7 x 2 - 5 = ☐

36 ÷ 9 x 3 = ☐

5 + 3 + 9 = ☐

132 ÷ 11 - 4 = ☐

13	14	3	8
12	20	27	2
75	7	11	51
19	58	30	74
40	48	64	4
95	9	17	31
22	46	21	83
33	5	16	90
15	87	92	10
6	1	23	108

©1996 Kelley Wingate Publications

37

CD-3724

Blankety- Blanks

Solve the problems below and write the answer in the box. On the blanket, shade in all the numbers that are in the answer boxes. The answers will make a pattern.

$5 \times 3 - 6 =$ ☐

$27 \div 3 + 4 =$ ☐

$3 \times 6 \div 9 =$ ☐

$120 \div 10 \times 11 =$ ☐

$32 \div 8 + 7 =$ ☐

$4 \times 3 - 9 =$ ☐

$2 \times 3 \times 5 =$ ☐

$21 \div 7 + 4 =$ ☐

$81 \div 9 + 8 =$ ☐

$3 \times 2 + 8 =$ ☐

$40 \div 5 + 8 =$ ☐

$56 \div 7 + 7 =$ ☐

$5 \times 8 \div 10 =$ ☐

$4 \times 9 \div 3 =$ ☐

$2 \times 9 \div 3 =$ ☐

$33 \div 3 - 6 =$ ☐

16	3	30	14
8	34	45	27
25	18	19	51
5	40	23	12
17	70	92	6
132	10	67	9
11	53	32	2
50	22	83	36
20	60	33	78
13	4	15	7

CD-3724

Name_____

Compare Squares

Compare the number sentences. Write <, >, or = in the square to make a true math statement. The first problem is done for you.

7 + 9 $\boxed{<}$ 5 x 4   50 ÷ 10 $\boxed{\phantom{<}}$ 15 - 9

56 ÷ 8 $\boxed{\phantom{<}}$ 15 - 7   4 x 7 $\boxed{\phantom{<}}$ 9 x 3

30 ÷ 3 $\boxed{\phantom{<}}$ 110 ÷ 10   21 ÷ 3 $\boxed{\phantom{<}}$ 14 - 6

17 - 8 $\boxed{\phantom{<}}$ 64 ÷ 8   54 ÷ 6 $\boxed{\phantom{<}}$ 13 - 5

15 - 9 $\boxed{\phantom{<}}$ 72 ÷ 9   7 + 6 $\boxed{\phantom{<}}$ 132 ÷ 12

132 ÷ 12 $\boxed{\phantom{<}}$ 3 x 4   17 - 9 $\boxed{\phantom{<}}$ 2 x 5

5 x 3 $\boxed{\phantom{<}}$ 6 + 7   35 ÷ 5 $\boxed{\phantom{<}}$ 42 ÷ 6

9 + 8 $\boxed{\phantom{<}}$ 4 x 5   6 x 4 $\boxed{\phantom{<}}$ 9 + 10

13 - 6 $\boxed{\phantom{<}}$ 48 ÷ 8   8 + 4 $\boxed{\phantom{<}}$ 81 ÷ 9

45 ÷ 5 $\boxed{\phantom{<}}$ 12 - 3   63 ÷ 7 $\boxed{\phantom{<}}$ 3 x 2

Compare Squares

Compare the number sentences. Write $<$, $>$, or $=$ in the square to make a true math statement. The first problem is done for you.

14 - 9 $\boxed{<}$ 42 ÷ 7   3 x 3 $\square$ 70 ÷ 7

5 + 7 $\square$ 108 ÷ 9   42 ÷ 6 $\square$ 15 - 9

63 ÷ 9 $\square$ 17 - 8   5 x 8 $\square$ 4 x 10

9 x 2 $\square$ 9 + 9   13 - 9 $\square$ 27 ÷ 9

25 ÷ 5 $\square$ 14 - 8   12 - 3 $\square$ 72 ÷ 6

14 - 6 $\square$ 72 ÷ 8   15 ÷ 5 $\square$ 20 ÷ 4

2 x 7 $\square$ 8 + 8   18 - 9 $\square$ 27 ÷ 3

21 ÷ 3 $\square$ 14 - 5   8 + 8 $\square$ 32 ÷ 2

18 - 9 $\square$ 24 ÷ 6   96 ÷ 8 $\square$ 36 ÷ 3

49 ÷ 7 $\square$ 16 - 9   2 x 7 $\square$ 6 + 7

                   CD-3724

Compare Squares

Compare the number sentences. Write <, >, or = in the square to make a true math statement. The first problem is done for you.

13 - 7 $\boxed{<}$ 49 ÷ 7   33 ÷ 11 $\square$ 3 x 3

21 ÷ 3 $\square$ 55 ÷ 11   12 ÷ 2 $\square$ 18 - 9

16 ÷ 4 $\square$ 2 x 3   2 x 9 $\square$ 7 + 9

9 + 9 $\square$ 3 x 6   9 + 2 $\square$ 44 ÷ 4

15 - 7 $\square$ 14 - 8   12 - 8 $\square$ 36 ÷ 12

14 - 9 $\square$ 20 ÷ 5   36 ÷ 6 $\square$ 4 x 3

3 x 8 $\square$ 5 x 5   5 x 6 $\square$ 7 x 4

96 ÷ 8 $\square$ 13 - 4   64 ÷ 8 $\square$ 32 ÷ 4

12 - 9 $\square$ 24 ÷ 12   16 - 9 $\square$ 45 ÷ 9

121 ÷ 11 $\square$ 16 - 8   8 x 6 $\square$ 16 - 7

Compare Squares

Compare the number sentences. Write <, >, or = in the square to make a true math statement. The first problem is done for you.

7 - 3 [<] 42 ÷ 7   16 - 7 [ ] 72 ÷ 9

6 x 6 [ ] 108 ÷ 9   110 ÷ 11 [ ] 14 - 8

5 x 3 [ ] 17 - 8   7 + 8 [ ] 3 x 6

81 ÷ 9 [ ] 9 + 9   17 - 9 [ ] 42 ÷ 7

35 ÷ 7 [ ] 14 - 8   9 x 4 [ ] 6 x 6

14 - 8 [ ] 72 ÷ 8   48 ÷ 8 [ ] 24 ÷ 8

144 ÷ 12 [ ] 8 + 8   13 - 9 [ ] 21 ÷ 7

12 ÷ 3 [ ] 14 - 5   6 x 3 [ ] 9 + 9

7 + 4 [ ] 24 ÷ 6   12 - 7 [ ] 45 ÷ 5

16 - 9 [ ] 14 - 7   4 x 3 [ ] 8 + 5

Name_____

Mystery Math

Look at the mystery number. Circle all math expressions in that row which equal the mystery number. The first problem is done for you.

Mystery Number	Math Expression			
3	28 ÷ 7	12 - 8	(27 ÷ 9)	(12 - 9)
11	4 + 7	77 ÷ 7	8 + 5	132 ÷ 12
4	15 - 9	32 ÷ 8	12 - 8	36 ÷ 6
6	15 - 8	24 ÷ 6	3 x 2	14 - 8
48	4 x 12	9 x 4	8 x 6	4 x 8
9	8 + 2	14 - 6	63 ÷ 7	17 - 8
12	96 ÷ 9	5 + 9	8 + 6	4 x 3
7	16 - 9	35 ÷ 7	49 ÷ 7	5 + 3
24	6 x 3	2 x 12	8 x 4	4 x 6

Mystery Math

Look at the mystery number. Circle all math expressions in that row which equal the mystery number. The first problem is done for you.

Mystery Number	Math Expression			
4	16 - 9	(12 - 8)	11 - 6	(16 ÷ 4)
11	5 + 7	6 + 5	121 ÷ 11	4 + 7
8	15 - 8	108 ÷ 12	17 - 9	96 ÷ 12
10	30 ÷ 5	7 + 3	110 ÷ 11	5 + 4
3	24 ÷ 6	11 - 8	36 ÷ 12	10 - 9
6	36 ÷ 9	15 - 9	72 ÷ 12	14 - 7
7	48 ÷ 6	13 - 6	16 - 9	42 ÷ 7
12	7 + 6	132 ÷ 12	60 ÷ 5	120 ÷ 10
5	35 ÷ 7	14 - 9	15 ÷ 5	30 ÷ 5

Name _____

Mega Mystery Math

Look at the mystery number. Circle all math expressions in that row which equal the mystery number. The first problem is done for you.

Mystery Number	Math Expression		
6	16 - 8 - 1	⟨15 ÷ 5 + 3⟩	⟨12 + 6 ÷ 3⟩
8	2 x 3 + 3	100 ÷ 10 - 2	18 ÷ 3 + 4
15	20 ÷ 4 x 3	17 - 8 + 6	4 x 2 + 8
17	81 ÷ 9 + 9	4 x 2 + 9	5 + 4 + 7
9	27 ÷ 3 + 3	14 - 7 + 2	60 ÷ 12 + 6
7	60 ÷ 5 - 5	5 x 3 - 8	14 - 8 + 1
3	42 ÷ 6 - 4	2 x 4 - 3	18 - 9 - 6
11	56 ÷ 8 + 4	63 ÷ 7 + 3	4 x 3 - 1
5	12 - 6 - 2	4 x 3 - 7	18 - 9 - 6

CD-3724

Mega Mystery Math

Look at the mystery number. Circle all math expressions in that row which equal the mystery number. The first problem is done for you.

Mystery Number	Math Expression		
9	$21 \div 7 + 3$	⟨$16 \div 4 + 5$⟩	⟨$4 \times 4 - 7$⟩
6	$18 - 9 - 2$	$24 \div 4 + 2$	$18 \div 6 + 3$
3	$24 \div 4 - 4$	$4 + 6 - 7$	$15 \div 3 - 2$
10	$45 \div 9 \times 2$	$5 \times 3 - 10$	$121 \div 11 - 1$
12	$21 \div 3 + 5$	$3 \times 2 \times 2$	$36 \div 6 \times 2$
8	$42 \div 7 + 3$	$5 + 6 - 4$	$18 \div 2 - 1$
2	$24 \div 6 - 2$	$13 - 5 + 3$	$4 \times 3 \div 6$
5	$27 \div 3 + 4$	$108 \div 12 - 4$	$2 \times 3 - 1$
7	$3 \times 5 - 8$	$12 - 8 + 4$	$63 \div 7 - 2$

   CD-3724

Magic Trail

Follow the trail by solving the math problems and find the magic number.

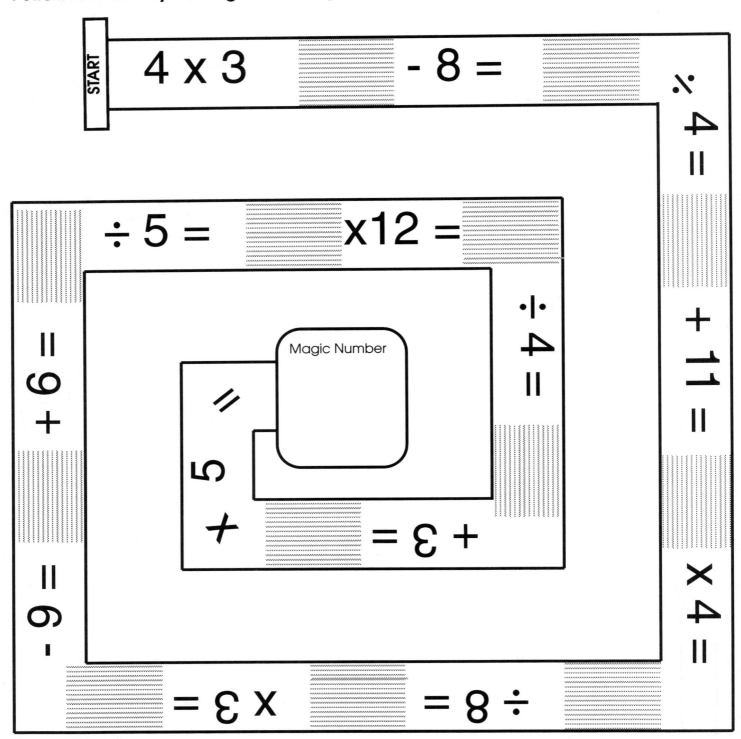

START

4 x 3 _____ - 8 = _____

÷ 4 =

+ 11 =

x 4 =

÷ 5 = _____ x12 = _____

÷ 4 =

+ 6 =

- 6 =

Magic Number

=

x 5 +

= 3 +

= 3 x _____ ÷ 8 =

MAGIC NUMBER _____

Magic Trail
Follow the trail by solving the math problems and find the magic number.

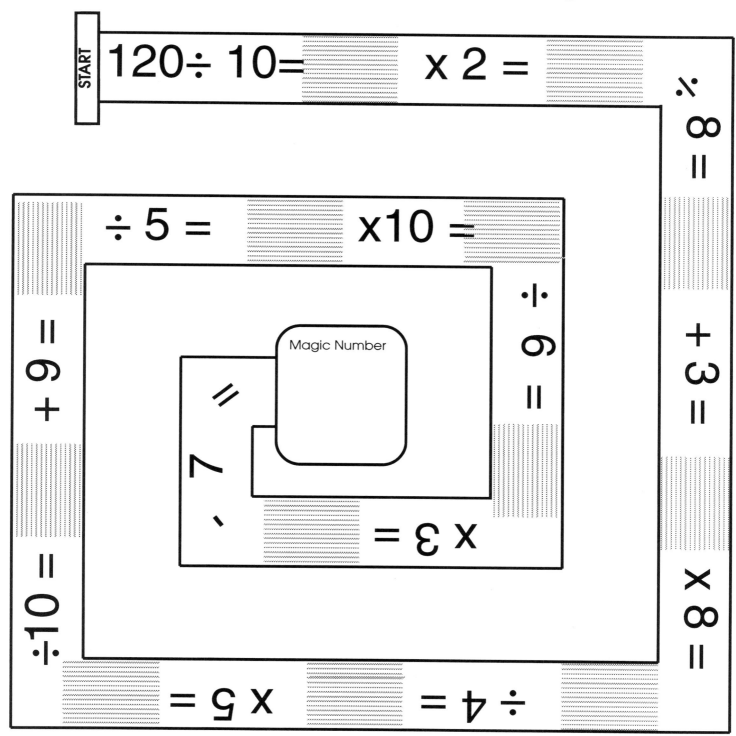

START
120 ÷ 10 = ___ x 2 = ___

÷ 8 =

÷ 5 = ___ x10 = ___

+ 3 =

+ 9 =

÷ 9 =

Magic Number

÷ 7 =

x 8 =

÷ 10 =

x 3 =

x 5 = ___ ÷ 4 = ___

MAGIC NUMBER _____

Place Space

Millions **,** Hundred Thousands | Ten Thousands | Thousands **,** Hundreds | Tens | Ones **.** Tenths | Hundredths | Thousandths

1. The number is:

| 7,320,196. 485 |

A. Name the digit in the tens place _____

B. Name the digit in the tenths place _____

C. Name the digit in the millions place _____

D. Name the digit in the ones place _____

E. In what place value is the digit "0"? _____

F. In what place value is the digit "4"? _____

G. In what place value is the digit "3"? _____

H. In what place value is the digit "5"? _____

2. The number is:

| 8,657.321 |

A. Name the digit in the hundreds place _____

B. Name the digit in the hundredths place _____

C. Name the digit in the thousands place _____

D. Name the digit in the tenths place _____

E. Name the number that is one hundred more _____

F. Name the number that is one thousand less _____

G. Name the number that is one hundredth less _____

H. Name the number that is one more _____

3. The number is:

| 7,320,196. 485 |

A. Name the digit in the millions place _____

B. Name the digit in the ones place _____

C. Name the digit in the thousandths place _____

D. Name the digit in the ten thousands place _____

E. Name the number that is ten thousand less _____

F. Name the number that is one thousandth more _____

G. Name the number that is one less _____

H. Name the number that is one million more _____

49

Place Space

Millions	,	Hundred Thousands	Ten Thousands	Thousands	,	Hundreds	Tens	Ones	.	Tenths	Hundredths	Thousandths

1. The number is: $\boxed{3,649,158.072}$

   A. Name the digit in the hundred thousands place _____

   B. Name the digit in the one thousands place _____

   C. Name the digit in the tens place _____

   D. Name the digit in the hundredths place _____

   E. In what place value is the digit "0"? _____

   F. In what place value is the digit "4"? _____

   G. In what place value is the digit "3"? _____

   H. In what place value is the digit "2"? _____

2. The number is: $\boxed{17,348.620}$

   A. Name the digit in the hundreds place _____

   B. Name the digit in the thousandths place _____

   C. Name the digit in the ten thousands place _____

   D. Name the digit in the tenths place _____

   E. Name the number that is one thousand more _____

   F. Name the number that is one tenth less _____

   G. Name the number that is ten more _____

   H. Name the number that is one hundred less _____

3. The number is: $\boxed{9,806,432.517}$

   A. Name the digit in the hundred thousands place _____

   B. Name the digit in the ones place _____

   C. Name the digit in the millions place _____

   D. Name the digit in the thousandths place _____

   E. Name the number that is ten thousand more _____

   F. Name the number that is one thousand less _____

   G. Name the number that is one hundred less _____

   H. Name the number that is one million less _____

Rounding Round-Up

To round any number, follow these simple rules:

Underline the place value you are rounding to.
Circle the digit to the right of the underlined digit.
If the circled number is 0, 1, 2, 3, or 4 the underlined digit stays the same.
If the circled number is 5, 6, 7, 8, or 9 the underlined digit goes up by 1.
The circled digit and all digits to the right become a zero.

Round to the nearest ten:	
364	366
3<u>6</u>4	3<u>6</u>6
3<u>6</u>4	3<u>6</u>6
360	370

1. Round these numbers to the nearest thousand:

A. 96,299 _____   B. 34,941 _____
C. 142,298 _____   D. 3,850 _____
E. 88,880 _____   F. 9,551 _____
G. 7,492 _____   H. 32,621 _____
I. 11,804 _____   J. 27,700 _____

2. Round these numbers to the nearest ten thousand:

A. 452,398 _____   B. 986,334 _____
C. 72,994 _____   D. 296,303 _____
E. 805,125 _____   F. 546,498 _____
G. 15,300 _____   H. 1,832,759 _____
I. 301,399 _____   J. 3,488,621 _____

3. Round these numbers to the nearest hundred thousand:

A. 43,645,117 _____   B. 416,667 _____
C. 6,285,136 _____   D. 3,821,501 _____
E. 19,915,324 _____   F. 2,137,148 _____
G. 743,698 _____   H. 292,531 _____
I. 461,433 _____   J. 3,554,712 _____

Rounding Round-Up

To round any number, follow these simple rules:

Underline the place value you are rounding to.
Circle the digit to the right of the underlined digit.
If the circled number is 0, 1, 2, 3, or 4 the underlined digit stays the same.
If the circled number is 5, 6, 7, 8, or 9 the underlined digit goes up by 1.
The circled digit and all digits to the right become a zero.

Round to the nearest ten:	
364	366
364	366
364	366
360	370

1. Round these numbers to the nearest tenth:

A. 56.324 _____   B. 41.12 _____

C. 18.76 _____   D. 307.685_____

E. 4.945_____   F. 10.852 _____

G. 132.75_____   H. 5.398 _____

I. 60.36 _____   J. 72.72_____

2. Round these numbers to the nearest hundredth:

A. 230.036_____   B. 155.872_____

C. 35.512 _____   D. 1,725.977 _____

E. 127.384_____   F. .296_____

G. 59.305 _____   H. 158.335_____

I. 82.361 _____   J. 12.765_____

3. Round these numbers to the nearest whole number (one):

A. 15.356 _____   B. 57.11_____

C. 37.84 _____   D. 18.9_____

E. 109.288 _____   F. 3.646_____

G. 722.510_____   H. 623.923_____

I. 78.41_____   J. 1,204.2_____

Rounding Round-Up

To round any number, follow these simple rules:

Underline the place value you are rounding to.
Circle the digit to the right of the underlined digit.
If the circled number is 0, 1, 2, 3, or 4 the underlined digit stays the same.
If the circled number is 5, 6, 7, 8, or 9 the underlined digit goes up by 1.
The circled digit and all digits to the right become a zero.

Round to the nearest ten:	
364	366
364	366
364	366
360	370

1. Round these numbers to the nearest thousandth:

A. 7.30452 _____ B. 12.00453 _____

C. 0.0151 _____ D. 7.98221 _____

E. 14.30649 _____ F. 27.01056 _____

G. 2.5428 _____ H. 124.5793 _____

I. 9.1342 _____ J. 36.4468 _____

2. Round these numbers to the nearest thousand:

A. 452,398 _____ B. 986,334 _____

C. 72,994 _____ D. 296,303 _____

E. 805,125 _____ F. 546,498 _____

G. 15,300 _____ H. 1,832,759 _____

I. 301,399 _____ J. 3,488,621 _____

3. Round these numbers to the nearest hundred thousand:

A. 624,677 _____ B. 552,542 _____

C. 285,453 _____ D. 1,712,899 _____

E. 681,398 _____ F. 9,107,600 _____

G. 519,601 _____ H. 243,377 _____

I. 154,429 _____ J. 3,094,192 _____

EXAMPLE:

Sometimes it is not important to have an exact answer. You can round the numbers and estimate what the answer will be.

$$47 \longrightarrow 50$$
$$\underline{+\ 52} \longrightarrow \underline{50}$$
$$100$$

Round to the highest level and estimate the answers:

137 →
+ 142 → _____

259 →
+ 619 → _____

838 →
- 266 → _____

38 →
+ 82 → _____

57 →
- 43 → _____

62 →
+ 49 → _____

73 →
- 47 → _____

93 →
+ 55 → _____

27 →
- 17 → _____

97 →
- 62 → _____

54 →
- 38 → _____

88 →
+ 36 → _____

856 →
+ 122 → _____

721 →
- 374 → _____

953 →
- 449 → _____

Name_____

Sometimes it is not important to have an exact answer. You can round the numbers and estimate what the answer will be.

EXAMPLE:

$47 \rightarrow 50$
$\underline{\times 52} \rightarrow \underline{50}$
2500

Round to the highest level and estimate the answers:

$629 \rightarrow$
$\underline{+ 384} \rightarrow \underline{}$

$718 \rightarrow$
$\underline{+ 949} \rightarrow \underline{}$

$608 \rightarrow$
$\underline{+ 284} \rightarrow \underline{}$

$824 \rightarrow$
$\underline{- 396} \rightarrow \underline{}$

$454 \rightarrow$
$\underline{- 243} \rightarrow \underline{}$

$733 \rightarrow$
$\underline{- 187} \rightarrow \underline{}$

$54 \rightarrow$
$\underline{\times 37} \rightarrow \underline{}$

$26 \rightarrow$
$\underline{\times 16} \rightarrow \underline{}$

$49 \rightarrow$
$\underline{\times 23} \rightarrow \underline{}$

$321 \rightarrow$
$\underline{\times 416} \rightarrow \underline{}$

$183 \rightarrow$
$\underline{\times 652} \rightarrow \underline{}$

$538 \rightarrow$
$\underline{\times 923} \rightarrow \underline{}$

$816 \rightarrow$
$\underline{\times 483} \rightarrow \underline{}$

$197 \rightarrow$
$\underline{\times 374} \rightarrow \underline{}$

$272 \rightarrow$
$\underline{\times 818} \rightarrow \underline{}$

CD-3724

Sometimes it is not important to have an exact answer. You can round the numbers and estimate what the answer will be.

EXAMPLE:

$453 \div 49 =$

↓   ↓

$500 \div 50 = 10$

Round to the highest level and estimate the answers:

$29 \div 11 =$

↓   ↓

$=$

$72 \div 11 =$

↓   ↓

$=$

$61 \div 9 =$

↓   ↓

$=$

$114 \div 14 =$

↓   ↓

$=$

$203 \div 22 =$

↓   ↓

$=$

$161 \div 37 =$

↓   ↓

$=$

$78 \div 15 =$

↓   ↓

$=$

$752 \div 159 =$

↓   ↓

$=$

$479 \div 51 =$

↓   ↓

$=$

$1,042 \div 451 =$

↓   ↓

$=$

$1,470 \div 19 =$

↓   ↓

$=$

$3,550 \div 375 =$

↓   ↓

$=$

Use the ten clues below to find the correct digits in this "magic" number.

____ , ____ ____ ____ , ____ ____ ____ . ____ ____ ____

1. The millions digit is the difference of 15 and 9.

   _____

2. The tenths digit is the quotient of 35 and 7.

   _____

3. The tens digit is the difference of 11 and 8.

   _____

4. The thousandths digit is the quotient of 20 and 5.

   _____

5. The hundreds digit is the difference of 8 and 7.

   _____

6. The hundred thousands digit is the product of 4 and 2.

   _____

7. The thousands digit is the sum of 1 and 1.

   _____

8. The ones digit is the quotient of 84 and 12.

   _____

9. The hundredths digit is the difference of 16 and 7.

   _____

10. The ten thousands digit is the product of 8 and 0.

   _____

Use the ten clues below to find the correct digits in this "magic" number.

___ , ___ ___ ___ , ___ ___ ___ . ___ ___ ___

1. The hundreds digit is the difference of 13 and 9.

   _____

2. The hundredths digit is the sum of 3 and 4.

   _____

3. The tenths digit is the quotient of 5 and 5.

   _____

4. The hundred thousands digit is the difference of 14 and 8.

   _____

5. The thousandths digit is the sum of 5 and 3.

   _____

6. The thousands digit is the quotient of 24 and 8.

   _____

7. The tens digit is the difference of 13 and 8.

   _____

8. The millions digit is the quotient of 22 and 11.

   _____

9. The ones digit is the difference of 2 and 2.

   _____

10. The ten thousands digit is the sum of 3 and 6.

   _____

Use the ten clues below to find the correct digits in this "magic" number.

___ , ___ ___ ___ , ___ ___ ___ . ___ ___ ___

1. The tenths digit is the quotient of 18 and 9.

   _____

2. The ten thousands digit is the difference of 12 and 7.

   _____

3. The millions digit is the sum of 7 and 2.

   _____

4. The hundredths digit is the quotient of 27 and 9.

   _____

5. The ones digit is the difference of 13 and 7.

   _____

6. The hundred thousands digit is the quotient of 32 and 8.

   _____

7. The tens digit is the difference of 3 and 3.

   _____

8. The thousands digit is the quotient of 84 and 12.

   _____

9. The hundreds digit is the difference of 13 and 5.

   _____

10. The thousandths digit is the quotient of 7 and 7.

   _____

Use the ten clues below to find the correct digits in this "magic" number.

___ , ___ ___ ___ , ___ ___ ___ . ___ ___ ___

1. The thousandths digit is the product of 11 and 0.

   _____

2. The hundred thousands digit is the quotient of 77 and 11.

   _____

3. The hundreds digit is the difference of 11 and 2.

   _____

4. The ones digit is the difference of 11 and 3.

   _____

5. The hundredths digit is the quotient of 22 and 11.

   _____

6. The millions digit is the product of 3 and 1.

   _____

7. The thousands digit is the difference of 11 and 7.

   _____

8. The tenths digit is the difference of 11 and 6.

   _____

9. The tens digit is the quotient of 66 and 11.

   _____

10. The ten thousands digit is the sum of 1 and 0.

   _____

Fraction - Ease

Fractions have **numerators** (top numbers) and **denominators** (bottom numbers).

When the denominators are the same you add or subtract the numerators. The denominator stays the same. Complete the problems below. The first one has been done for you.

$\dfrac{1}{3} + \dfrac{1}{3} = \boxed{\dfrac{2}{3}}$

$\dfrac{2}{5} - \dfrac{1}{5} = \boxed{}$

$\dfrac{2}{7} + \dfrac{3}{7} = \boxed{}$

$\dfrac{6}{11} - \dfrac{4}{11} = \boxed{}$

$\dfrac{1}{8} + \dfrac{1}{8} = \boxed{}$

$\dfrac{2}{9} + \dfrac{6}{9} = \boxed{}$

$\dfrac{5}{9} - \dfrac{3}{9} = \boxed{}$

$\dfrac{2}{5} + \dfrac{2}{5} = \boxed{}$

$\dfrac{6}{7} - \dfrac{3}{7} = \boxed{}$

$\dfrac{3}{4} - \dfrac{1}{4} = \boxed{}$

$\dfrac{1}{2} - \dfrac{1}{2} = \boxed{}$

$\dfrac{2}{6} + \dfrac{3}{6} = \boxed{}$

$\dfrac{1}{5} + \dfrac{3}{5} = \boxed{}$

$\dfrac{5}{7} - \dfrac{1}{7} = \boxed{}$

$\dfrac{8}{9} - \dfrac{7}{9} = \boxed{}$

$\dfrac{1}{10} + \dfrac{3}{10} = \boxed{}$

$\dfrac{5}{6} - \dfrac{4}{6} = \boxed{}$

$\dfrac{5}{8} + \dfrac{2}{8} = \boxed{}$

Decimal Dimensions

To add or subtract decimals, line up the decimal points and fill in any blank spaces with a zero. Add or subtract as usual. Do not forget to bring the decimal point straight down into your answer.

EXAMPLES:

1.6	**01.60**	01.60		24.8	24.**80**	24.80
+ 33.45	+ 33.45	+ 33.45		- 13.75	- 13.75	- 13.75
		35.05				**11.05**

1)
$$\begin{array}{r} 6.25 \\ 14.0 \\ +\ 36.75 \\ \hline \end{array}$$

2)
$$\begin{array}{r} 4.3 \\ -\ 2.007 \\ \hline \end{array}$$

3)
$$\begin{array}{r} 91.8 \\ -\ 12.17 \\ \hline \end{array}$$

4)
$$\begin{array}{r} 5.5 \\ 3.03 \\ +\ 0.004 \\ \hline \end{array}$$

5)
$$\begin{array}{r} 116.09 \\ -\ 97.387 \\ \hline \end{array}$$

6)
$$\begin{array}{r} 19.82 \\ -\ 7.002 \\ \hline \end{array}$$

7)
$$\begin{array}{r} 49.7 \\ -\ 39.95 \\ \hline \end{array}$$

8)
$$\begin{array}{r} 5.2 \\ -\ 3.84 \\ \hline \end{array}$$

9)
$$\begin{array}{r} 0.51 \\ +\ 3.60 \\ +14.00 \\ \hline \end{array}$$

10)
$$\begin{array}{r} 17.0 \\ 3.9 \\ +\ 1.75 \\ \hline \end{array}$$

11)
$$\begin{array}{r} 440.01 \\ -\ 289.607 \\ \hline \end{array}$$

12)
$$\begin{array}{r} 40.0 \\ -\ 38.81 \\ \hline \end{array}$$

Decimal Dimensions

To multiply decimals: 1) Multiply as usual; 2) In the final product, move the decimal once to the left for every decimal place in the original two multipliers (numbers multiplied together)

EXAMPLES:

1.5	1.5	1 place
x 0.3	x 0.3	1 place
045	0.45	2 places

8.3	8.3	1 place
x .002	x .002	3 places
00166	0.0166	4 places

1)　　　0.4
　　　x 0.8

2)　　　1.6
　　　x　 3

3)　　　21.2
　　　x　 3.3

4)　　　7.01
　　　x　 0.4

5)　　　9.8
　　　x 0.3

6)　　　.004
　　　x　　.5

7)　　　1.23
　　　x　 5.4

8)　　　28.2
　　　x　　.4

9)　　　4.97
　　　x　　6

10)　　　5.8
　　　x　　.3

11)　　　2.5
　　　x 3.7

12)　　　5.08
　　　x　 7.2

CD-3724

Decimal Dimensions

To multiply decimals: 1)Multiply as usual; 2) In the final product, move the decimal once to the left for every decimal place in the original two multipliers (numbers multiplied together)

EXAMPLES:

1.5	1.5	1 place
x 0.3	x 0.3	1 place
045	0.45	2 places

8.3	8.3	1 place
x .002	x .002	3 places
00166	0.0166	4 places

1) 1.8 x 4	2) 7.3 x 8	3) 5.05 x 3
4) 4.43 x 7	5) 4.8 x 2.3	6) 5.2 x 8.9
7) 27.4 x 1.7	8) 38.4 x 3.2	9) .84 x 6.7
10) 3.01 x 4.9	11) 5.003 x 1.4	12) .006 x .2

Measure Sense
Standard

Length
1 inch (in)
12 ins = 1 foot (ft)
3 ft = 1 yard (yd)
5,280 ft = 1 mile (mi)

1. length of a football field	2. height of a doorway	3. width of a math book
100 yd. 100 mi.	7 in. 7 ft.	8 in. 8 ft.
4. height of a desk	5. length of a swimming pool	6. width of your wrist
2 in. 2 ft.	36 ft. 36 mi.	3 in. 3 ft.
7. length of your foot	8. length of a car	9. height of a fence
7 in. 7 ft.	100 in. 100 ft.	4 in. 4 ft.
10. your house to the corner	11. height of a basketball player	12. height of a house
50 yd. 50 mi.	7 ft. 7 yd.	20 ft. 20 yd.

Measure Sense
Metric

Length	
Millimeter (mm): thickness of a coin	**10 mm. = 1 cm.**
Centimeter (cm): one fingernail	**10 cm. = 1 dm.**
Decimeter (dm): height of a small can	**10 dm. = 1 m.**
Meter (m): lenth of a baseball bat	**1000 m. = 1 km.**
Kilometer (km): about 2/3 of a mile	

1. height of a wall in your home 3 mm.　　　3 m.	2. piece of notebook paper 15 cm.　　　15 dm.	3. thickness of an envelope 1 mm.　　　1 cm.
4. length of a football 3 cm.　　　3 dm.	5. width of a dictionary 7 mm.　　　7 cm.	6. height of a fence 2 cm.　　　2 m.
7. liter of soda 30 cm.　　　30 m.	8. height of a chair 1 dm.　　　1 m.	9. from one city to another 40 m.　　　40 km.
10. your house to your school 2 m.　　　2 km.	11. width of your desk 4 mm.　　　4 dm.	12. thickness of a computer disk 1 mm.　　　1 dm.

Conversion Excursion
Standard

Use the conversion chart to finish the statements below.

> **Length**
> 12 in. = 1 ft.
> 3 ft. = 1 yd.
> 5,280 ft. = 1 mi.
> 1,760 yd. = 1 mi.

1 ft. = ＿＿＿＿＿＿＿＿＿ in.

1 yd. = ＿＿＿＿＿＿＿＿＿ in.

1 mi. = ＿＿＿＿＿＿＿＿＿ yd.

24 in. = ＿＿＿＿＿＿＿＿＿ ft.

3 mi. = ＿＿＿＿＿＿＿＿＿ ft.

10,560 ft. = ＿＿＿＿＿＿＿ mi.

12 ft. = ＿＿＿＿＿＿＿＿＿ yd.

48 in. = ＿＿＿＿＿＿＿＿＿ ft.

60 in. = ＿＿＿＿＿＿＿＿＿ ft.

36 ft. = ＿＿＿＿＿＿＿＿＿ yd.

5 yd. = ＿＿＿＿＿＿＿＿＿ ft.

6 yd. = ＿＿＿＿＿＿＿＿＿ ft.

10 ft. = ＿＿＿＿＿＿＿＿＿ in.

36 in. = ＿＿＿＿＿＿＿＿＿ ft.

Measure Sense
Standard

Weight
16 ounces (oz.) = 1 pound (lb.)
2,000 lb. = 1 ton (tn.)
1 slice of bread weighs about 1 oz.
1 loaf of bread weighs about 1 lb.

1. a tennis shoe	2. a handful of popcorn	3. a grown man
2 oz. 2 lb.	3 oz. 3 lb.	220 oz. 220 lb.
4. a baseball	5. your math book	6. a loaf of bread
6 oz. 6 lb.	4 oz. 4 lb.	1 oz. 1 lb.
7. a pencil	8. a baby	9. a chair
1 oz. 1 lb.	7 oz. 7 lb.	10 oz. 10 lbs.
10. an elephant	11. a car	12. an umbrella
5 lb. 5 tn.	2 lb. 2 tn.	1 lb. 1 tn.

Conversion Excursion
Standard

Name _____

Use the conversion chart to finish the statements below.

```
┌─────────────────────────────────┐
│        Weight                   │
│                                 │
│     16 oz.  =  1 lb.            │
│    2,000 lb.  = 1 ton           │
│                                 │
└─────────────────────────────────┘
```

4 ton = _____ lb.

4,000 lb. = _____ ton

6 lb. = _____ oz.

5 ton = _____ lb.

1600 oz. = _____ lb.

32 oz. = _____ lb.

3 lb. = _____ oz.

160 oz. = _____ lb.

80 oz. = _____ lb.

10,000 lb. = _____ ton

10 ton = _____ lb.

12,000 lb. = _____ ton

48 oz. = _____ lb.

6 lb. = _____ oz.

CD-3724

Measure Sense
Metric

┌─────────────────────────────────────┐
Weight
1,000 grams (g) = 1 kilogram (kg)

Gram (g): weight of of one grape
Kilogram (kg): weight of a bunch
of bananas
└─────────────────────────────────────┘

1. a feather	2. a car	3. a bag of potatoes
1 g. 1 kg.	1,000 g. 1,000 kg.	2 g. 2 kg.
4. a child	5. a dinner plate	6. a table
30 g. 30 kg.	500 g. 500 kg.	40 g. 40 kg.
7. a slice of cheese	8. one orange	9. a basketball
3 g. 3 kg.	100 g. 100 kg.	800 g. 800 kg.
10. a photograph	11. a shoe	12. a desk
1 g. 1 kg.	700 g. 700 kg.	5 g. 5 kg.

Measure Sense
Standard

Capacity
1 cup (c)
2 c = 1 pint (pt)
2 pts = 1 quart (qt)
4 qts = 1 gallon (gal)

1. a bowl of soup	2. a tea kettle	3. a bathtub full of water
2 c. 2 qt.	4 c. 4 gal.	30 pt. 30 gal.
4. a glass of soda	5. a fish tank	6. a pitcher of tea
1 c. 1 gal.	10 c. 10 gal.	2 c. 2 qt.
7. a can of tomato sauce	8. gas to fill a car	9. milk for the week
1 pt. 1 gal.	10 qt. 10 gal.	1 c. 1 gal.
10. a swimming pool	11. water to fill a vase	12. water to fill a bucket
400 pt. 400 gal.	3 c. 3 gal.	2 pt. 2 gal.

Measure Sense
Metric

Capacity
1,000 mL. = 1 liter (L)
Milliliter (ml) = eye dropper full
Liter (l) = half of a 2 liter bottle of soda

1. tablespoon of milk	2. a pail of water	3. a bowl of soup
10 mL. 100 mL.	3 mL. 3 L.	45 mL. 450 mL.
4. a cup of tea	5. a bathtub full of water	6. a raindrop
150 mL. 150 L.	200 mL. 200 L.	1 mL. 1 L.
7. a swimming pool	8. a thimble of water	9. a carton of milk from the cafeteria
100 L. 1,000 L.	15 mL. 15 L.	10 mL. 100 mL.
10. a bottle of soda	11. gas to fill a car	12. water for a fish tank
2 mL. 2 L.	1 L. 15 L.	15 mL. 15 L.

Conversion Excursion
Standard

Use the conversion chart to finish the statements below.

<u>Capacity</u>

2 c. = 1 pt
2 pt. = 1 qt.
4 qt. = 1 gal

8 pt. = _____ qt.

20 c. = _____ pt.

4 qt. = _____ pt.

32 qt. = _____ gal.

12 pt. = _____ qt.

16 qt. = _____ gal.

32 c. = _____ qt.

6 gal. = _____ qt.

4 pt. = _____ c.

14 c. = _____ pt.

10 c. = _____ pt.

4 gal. = _____ pt.

5 qt. = _____ c.

3 qt. = _____ pt.

Name _____

Measure Sense
Celsius (C) and Fahrenheit (F)

Temperature		
Fahrenheit (F) **Celsius** (C)		
Water freezes	32 ° F	0 ° C
Water boils	212 ° F	100 ° C

1. a cup of hot chocolate 90° C. 90° F.	2. a snowy day 20° C. 20° F.	3. a warm day 25° C. 100° C.
4. ice 20° C. 0° C.	5. a hot oven 50° F. 400° F.	6. a cold drink 40° C. 40° F.
7. a hot day 50° C. 50° F.	8. a comfortable room 68° C. 68° F.	9. boiling water 0° C. 100° C.
10. hot water for coffee or tea 100° C. 100° F.	11. ice cream 32° C. 32° F.	12. normal body temperature 98° C. 98° F.

Measure Sense
Time

Time
60 seconds (sec.) = 1 minute (min.)
60 min. = 1 hour (hr.)
24 hr. = 1 day
7 days = 1 week (wk)
365 days = 52 wk. = 1 year (yr.)

1. read a 300 page book	2. take a short test	3. take a shower
1 min. 1 wk.	15 sec. 15 min.	10 min. 10 days
4. sleep at night	5. watch a movie	6. clean your room
8 min. 8 hr.	2 min. 2 hr.	20 min. 20 yr.
7. eat dinner	8. wash your car	9. say your name
20 sec. 20 min.	15 sec. 15 min.	2 sec. 2 min.
10. grow from a seed to a tree	11. eat three meals	12. grow a plant
6 wk. 6 yr.	1 day 1 wk.	5 hr. 5 wk.

Conversion Excursion
Standard

Use the conversion chart to finish the statements below.

```
          Time
   60 sec. = 1 min.
   60 min.  = 1 hr.
   24 hr.  = 1 day
   7 days = 1 wk
365 days = 52 wk. = 1 yr.
```

4 min. = _____ sec.

120 sec. = _____ min.

3 days = _____ hr.

15 hr. = _____ min.

600 min. = _____ hr.

2 yr. = _____ wk.

8 hr. = _____ min.

7 wk. = _____ days

48 hr. = _____ days

1 yr. = _____ wk.

30 min. = _____ sec.

365 days = _____ wk.

9 days = _____ hr.

21 days = _____ wk.

CD-3724

Conversion Excursion

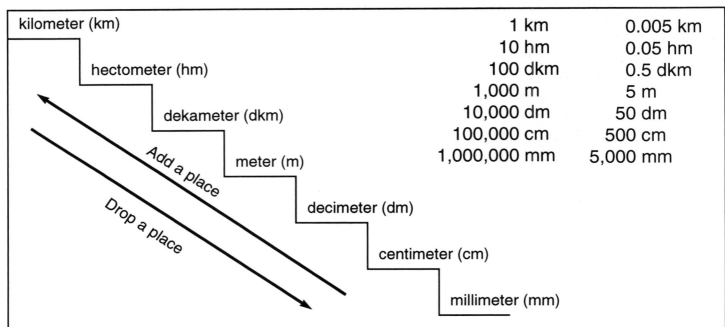

	1 km	0.005 km
	10 hm	0.05 hm
	100 dkm	0.5 dkm
	1,000 m	5 m
	10,000 dm	50 dm
	100,000 cm	500 cm
	1,000,000 mm	5,000 mm

When converting from one metric unit to another: Move the decimal once to the right for each step down (the number gets larger). Move the decimal once to the left for each step up (the number gets smaller).

Use the metric stairs to convert these measurements from one unit to another.

6 km = _____ m   400 mm = _____ dm

30 cm = _____ dm   50 km = _____ cm

7,000 mm = _____ m   20 m = _____ dm

800 cm = _____ mm   40 m = _____ cm

750,000 m = _____ km   14,000 mm = _____ cm

200 dkm = _____ cm   576.4 m = _____ cm

13 hm = _____ m   1.37 km = _____ dm

Conversion Excursion

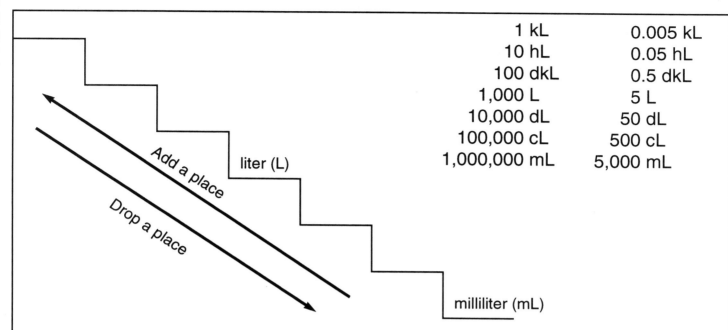

1 kL   0.005 kL
10 hL   0.05 hL
100 dkL   0.5 dkL
1,000 L   5 L
10,000 dL   50 dL
100,000 cL   500 cL
1,000,000 mL 5,000 mL

Add a place

Drop a place

liter (L)

milliliter (mL)

When converting from one metric unit to another: Move the decimal once to the right for each step down (the number gets larger). Move the decimal once to the left for each step up (the number gets smaller).

Fill in the metric stairs and use them to convert these measurements.

4 mL = _____ L   5,723 mL = _____ dL

2 L = _____ dL   5.6 kL = _____ cL

18,000 mL = _____ hL   69 L = _____ dL

650 cL = _____ mL   292 L = _____ cL

174,000 dL = _____ kL   12,300 mL = _____ dL

325 dkL = _____ cL   43.24 L = _____ mL

48.7 hL = _____ L   16.8 kL = _____ L

Name _____

Conversion Excursion

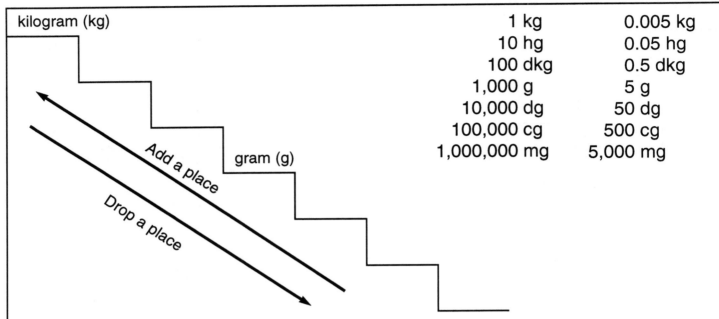

kilogram (kg)

1 kg	0.005 kg
10 hg	0.05 hg
100 dkg	0.5 dkg
1,000 g	5 g
10,000 dg	50 dg
100,000 cg	500 cg
1,000,000 mg	5,000 mg

Add a place

gram (g)

Drop a place

When converting from one metric unit to another: Move the decimal once to the right for each step down (the number gets larger). Move the decimal once to the left for each step up (the number gets smaller).

Fill in the metric stairs and use them to convert these measurements.

3 hg = _____ cg   887 mg = _____ dg

75 dkg = _____ dg   73 kg = _____ cg

4,120 mg = _____ g   316 g = _____ dg

99 dg = _____ mg   123 g = _____ cg

58,342 mg = _____ hg   256,000 mg = _____ cg

4,740 g = _____ kg   33.5 g = _____ cg

27.4 hg = _____ g   87.5 kg = _____ dg

                   CD-3724

Conversion Excursion
Mixed Practice

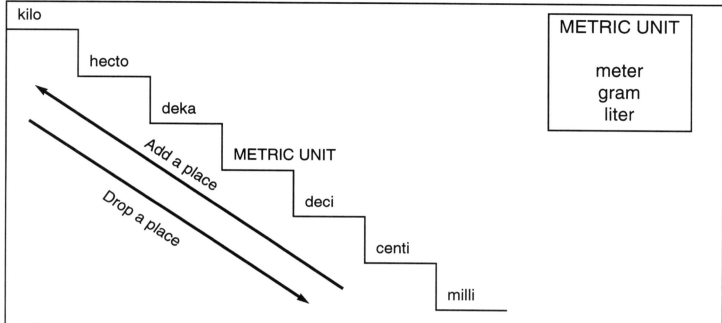

METRIC UNIT
meter
gram
liter

When converting from one metric unit to another: Move the decimal once to the right for each step down (the number gets larger). Move the decimal once to the left for each step up (the number gets smaller).

Use the metric stairs to convert these measurements from one unit to another.

50 dm = _____ mm   15 mm = _____ hm

5 km = _____ dm   7 m = _____ mm

4,600 mL = _____ dkL   28 L = _____ hL

9,300 cL = _____ dL   647 mL = _____ dL

900,000 mL = _____ kL   158,000 mg = _____ cg

80 dkg = _____ cg   35 g = _____ cg

61 hg = _____ g   .448 kg = _____ dg

Conversion Excursion

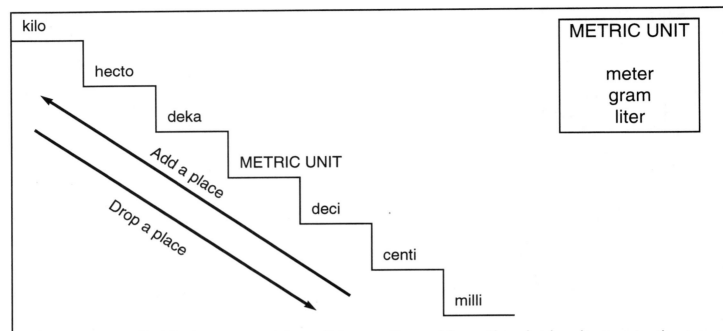

When converting from one metric unit to another: Move the decimal once to the right for each step down (the number gets larger). Move the decimal once to the left for each step up (the number gets smaller).

Use the metric stairs to convert these measurements from one unit to another.

25,000mm = _____ cm   3,000 g = _____ kg

400 L = _____ hL   6,000 cm = _____ km

9 km = _____ m   30 dL = _____ L

24 kg = _____ cg   741 m = _____ cm

1,800 dm = _____ mm   21,300 mg = _____ g

70 dkL = _____ L   .65 m = _____ cm

58 hg = _____ dg   .045 kL = _____ cL

Words Into Math

FOOD	VOTES
Tacos	★★★★★★
Chicken Nuggets	★★★★
Pizza	★★★★★★★★★★★★
Hot Dog	★★★★★★
Hamburger	★★★★★★★★
Spaghetti	★★★

Jessica and Erica did a survey for a math project. They charted the favorite cafeteria foods of their whole 4th grade math class. Each star stands for one vote.

1. Which food do the students like best?

_____

2. How many students are in this 4th grade math class?

_____

3. How many more children voted for tacos than spaghetti?

_____

4. How many students voted for pizza or hotdogs?

_____

5. Which food did Erica vote for?

_____

6. How many more students voted for pizza than spaghetti?

_____

7. What class gets to vote next week?

8. Which food got the fewest votes?

_____

9. Which two foods got the same number of votes?

_____

10. How many students wanted to add red beans and rice to the menu?

_____

Words Into Math

Daily Schedule	TIME
Homeroom	8:00 - 8:15
1st Period	8:15 - 9:15
2nd Period	9:20 - 10:20
3rd Period	10:25 - 11:25
Lunch	11:25 - 12:00
4th Period	12:00 - 1:00
5th Period	1:05 - 2:05
6th Period	2:10 - 3:10

Charles High School has this daily schedule. It names all of the class periods and gives the time each one starts and ends. Use this schedule to answer the questions below.

1. How long is each class period?

_____

2. If Jimmy gets to class at 8:30, how much of 1st period did he miss?

_____

3. What class does Susan have 4th Period?

_____

4. Kaitlin's mom picked her up at 2:00. During what period did she leave?

_____

5. Jamal stays in Mr. Woods room for 2nd and 3rd period. How long is he in that room?

_____

6. What period does Ms. Smith teach gym?

_____

7. How long is the lunch period?

_____

8. Which period begins at 9:20?

_____

9. What event happens between 8 and 8:15 each morning?

_____

10. If school is over after 6th period, what time do the students get out?

_____

Words Into Math

Read the paragraph carefully then answer the questions.

ADMIT ONE
ZOO TICKET

ADMIT ONE
BUS FARE

Mrs. Gros's 4th grade class is going on a field trip to the zoo. There are 25 students in her class. Each child must pay $2.00 to get in the zoo and $1.25 for the bus fare. The class left the school at 9:00 and returned at 2:20.

1. What was the total each student paid to go on this trip?

2. How much money did Mrs. Gros collect from the class all together?

3. One parent for every 5 students came along on the trip. How many parents came?

4. Did Daniel like the sharks or the polar bears better?

5. How long was the class gone from school?

6. How much did the zoo collect for all 25 students?

7. How much more did it cost each child to get in the zoo than to ride the bus?

8. Did the students behave nicely on this trip?

Name _____

Words Into Math

At Mr. Lopez's sporting goods store, these items are on sale this week.

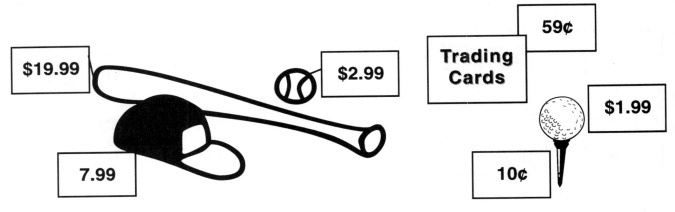

$19.99

$2.99

Trading Cards

59¢

$1.99

7.99

10¢

1. Javier bought a new bat and three balls. How much did he spend?

_____

2. Eric has $2.00. How many packs of trading cards can he buy?

_____

3. What is the cost of a bat and a golf ball together?

_____

4. Kara has $28.00. Can she buy the bat and a ball or just one of them?

_____

5. How many baseballs can Carlos get for his $10.00?

_____

6. Does the store have Jenny's size shoes?

_____

7. Which costs more: a bat and cards, or a ball and a hat?

8. How many more hits did Matthew get after he used his new bat?

_____

9. Which costs more: one hat, two baseballs, or ten packs of trading cards?

_____

10. Round each item to the nearest dollar. How much would it cost to get one of everything?

_____

Words Into Math

Mrs. Jacob put up a chart to show how many canned goods her students collected for the food drive. Use the chart to answer the questions.

Students Canned goods					
Nicole	♥	Halie	♥♥♥	Kara	♥♥♥♥
Miriam	♥♥	Josh	♥♥	Eric	♥♥♥❣
Javier	♥♥♥♥	Catherine	♥	Kaitlin	♥♥❣
Charlie	❣	Jimmy	♥❣	Matt	♥♥♥♥♥♥❣
Tina	❣	Ashley	♥♥	Lindsey	♥
Tommy	♥	Brandon	♥❣	Jesse	♥❣

❣ = 1 can   ♥ = 2 cans

1. Which student brought in the most cans?

_____

2. Which students brought in the fewest cans?

_____

3. How many students brought in three or less cans?

_____

4. How many students are in Mrs. Jacob's class?

_____

5. Is Jesse a boy or a girl?

_____

6. How many cans did the class collect all together?

_____

7. How many more cans did Matt bring than Josh?

_____

8. How many cans did Kaitlin, Kara, and Catherine bring in together?

_____

9. What kind of food was in Charlie's can?

_____

10. Who brought in more: Eric and Ashley, or Javier and Miriam?

_____

Skills Evaluation

Choose the best answer to these review questions. Circle the correct answer.

1. Add: 43,442 + 24,758 A. 67,100 B. 67,390 C. 68,190 D. 68,200	2. Subtract: 3,047 - 1,869 A. 1,078 B. 1,178 C. 2,822 D. 2,812
3. Subtract: 85,040 - 43,639 A. 41,419 B. 42,679 C. 41,401 D. 42,411	4. Multiply: 392 x 47 A. 18,424 B. 14,862 C. 27,394 D. 18,433
5. Which of these expressions does <u>not</u> equal 9? A. 17 - 8 B. 63 ÷ 7 C. 40 ÷ 5 D. 14 - 5	6. Which of these expressions does <u>not</u> equal 6? A. 24 ÷ 6 B. 15 - 9 C. 72 ÷ 12 D. 3 x 2
7. Solve: 3 x 8 ÷ 6 = ☐ A. 6 B. 5 C. 4 D. 3	8. Solve: 6 x 6 ÷ 9 = ☐ A. 3 B. 4 C. 5 D. 6
9. Divide: 324 ÷ 6 = ☐ A. 204 B. 64 C. 53 D. 54	10. Divide: 504 ÷ 8 = ☐ A. 63 B. 86 C. 74 D. 52

Skills Evaluation

Choose the best answer to these review questions. Circle the correct answer.

1. Add: $\dfrac{1}{4} + \dfrac{1}{3}$	2. Add: $\dfrac{2}{5} + \dfrac{3}{6}$
A. 2/7 B. 2/4 C. 7/12 D. 4/3	A. 5/11 B. 5/14 C. 5/30 D. 27/30
3. The number is 3,157,246. The 3 is in which place? A. hundreds B. thousands C. thousandths D. millions	4. The number is 403,521.78. The 8 is in which place? A. ones B. tenths C. thousandths D. hundredths
5. What is the least common denominator for 5/6 and 2/9 ? A. 54 B. 18 C. 36 D. 9	6. What is the least common denominator for 4/5 and 7/8 ? A. 16 B. 32 C. 40 D. 45
7. Round to the nearest tenth: 33.765 A. 30 B. 34 C. 33.7 D. 33.8	8. Round to the nearest hundred: 32,172 A. 32,170 B. 32, 200 C. 32,000 D. 30,000
9. Round to the nearest ten thousand: 154,998 A. 154,000 B. 155,000 C. 150,000 D. 160,000	10. Round to the nearest million: 75,523,479 A. 75,000,000 B. 76,000,000 C. 70,000,000 D. 80,000,000

Skills Evaluation

Choose the best answer to these review questions. Circle the correct answer.

1. Tell if the expression is <, >, or =.

$72 \div 8 \ \square \ 15 - 9$

A. >   B. <

C. =

2. Tell if the expression is <, >, or =.

$12 \times 3 \ \square \ 4 \times 9$

A. >   B. <

C. =

3. Tell if the expression is <, >, or =.

$344 \div 8 \ \square \ 5 \times 9$

A. >   B. <

C. =

4. Tell if the expression is <, >, or =.

$27 + 31 \ \square \ 6 \times 9$

A. >   B. <

C. =

5. Add: $1.06 + 3.9 + 5$

A. 8.69   B. 9.96

C. 1.50   D. 4.01

6. Subtract: $5 - 3.789$

A. 3.784   B. 1.211

C. 3.794   D. 1.221

7. Multiply: $\begin{array}{r} 452 \\ \times \ 7 \\ \hline \end{array}$

A. 2,852   B. 2,862

C. 3,064   D. 3,164

8. Multiply: $\begin{array}{r} 1640 \\ \times \ 62 \\ \hline \end{array}$

A. 10,168   B. 3,280

C. 101,680   D. 98,480

9. Multiply: $\begin{array}{r} 4.22 \\ \times \ 3.6 \\ \hline \end{array}$

A. 15.192   B. 1.5192

C. 151.92   D. 1519.2

10. Multiply: $\begin{array}{r} 35.8 \\ \times \ .034 \\ \hline \end{array}$

A. 121.7200   B. 12.1720

C. 1.2172   D. .12127

Skills Evaluation

Choose the best answer to these review questions. Circle the correct answer.

1. Divide: $43 \div 7$ A. 6 B. 6 r 1 C. 6 r 3 D. 6 r 2	2. Divide: $65 \div 8$ A. 8 B. 8 r 1 C. 8 r 3 D. 8 r 5
3. Divide: $303 \div 4$ A. 76 B. 75 r 3 C. 75 r 2 D. 75	4. Divide: $247 \div 5$ A. 9 r 2 B. 39 r 2 C. 49 r 1 D. 49 r 2
5. Find the equal measurement: 5,000 mL. = A. 500 L. B. 50 L. C. 5 L. D. 0.5 L	6. Find the equal measurement: 6 c. = A. 1 qt. B. 2 pt. C. 3 pt. D. 1 gal.
7. Find the equal measurement: 48 hr. = A. 1 day B. 96 mins. C. 2 days D. 1 wk.	8. Find the equal measurement: 60 cm. = A. 6 mm. B. 60 mm. C. 600 mm. D. 6,000 mm.
9. Find the equal measurement: 5,000 dkm = A. 500 km B. 500,000 dm C. 50 m D. 5 cm	10. Find the equal measurement: 240 cg = A. 24 mm B. 24,00 g C. 0.24 dkg D. 2,400 dg

Math Whiz!

_____

receives this award for

_____

Keep up the great work!

_____ _____

signed   date

Dazzling Division!

_____

receives this award for

_____

Keep up the dazzling work!

_____ _____

signed   date

CD-3724

Multiplication Award

_____

receives this award for

_____

Keep up the great work!

_____ _____
signed   date

Place Value Superstar!

_____

is a Place Value Superstar!

_____

You are terrific!

_____ _____
signed   date

CD-3724

Answer Key

Name _____ Skill: Regrouping Addition Practice

RAP

Add the following problems. Do not forget to regroup when necessary.

Name _____ Skill: Regrouping Addition Practice

RAP

Add the following problems. Do not forget to regroup when necessary.

68 +34 = **102**	56 +97 = **153**	83 +58 = **141**	77 +84 = **161**	48 +92 = **140**
184 +717 = **901**	556 +385 = **941**	964 +328 = **1292**	829 +347 = **1176**	726 +478 = **1204**
851 +579 = **1430**	365 +498 = **863**	982 +469 = **1451**	735 +695 = **1430**	578 +632 = **1210**
3,842 +6,787 = **10,629**	5,096 +2,988 = **8,084**	8,147 +1,676 = **9,823**	7,865 +4,348 = **12,213**	8,907 +4,537 = **13,444**

6,469 + 4,153 = **10,622**   3,967 + 8,946 = **12,913**

8,355 + 4,987 = **13,342**   3,234 + 7,966 = **11,200**

5,463 + 7,549 = **13,012**   6,516 + 6,488 = **13,004**

6,308 + 8,942 = **15,250**   3,377 + 9,642 = **13,019**

3,845 + 8,628 = **12,473**   4,856 + 5,865 = **10,721**

Name _____ Skill: Regrouping Addition Practice

RAP

Add the following problems. Do not forget to regroup when necessary.

37 48 +92 = **177**	74 65 +38 = **177**	69 78 +36 = **183**	55 87 +32 = **174**	88 37 +86 = **211**
127 192 +253 = **572**	146 523 +978 = **1647**	847 453 +796 = **2096**	388 947 +365 = **1700**	982 637 +256 = **1875**
576 834 +429 = **1839**	822 739 +245 = **1806**	157 495 +633 = **1285**	521 709 +483 = **1713**	712 516 +294 = **1522**

469 + 153 + 333 = **965**   967 + 946 + 254 = **2167**

355 + 987 + 613 = **1955**   234 + 966 + 863 = **2063**

463 + 549 + 243 = **1255**   516 + 488 + 119 = **1123**

308 + 942 + 876 = **2126**   377 + 642 + 629 = **1648**

845 + 628 + 458 = **1931**   856 + 865 + 138 = **1859**

Name _____ Skill: Regrouping Addition Practice

RAP

Add the following problems. Do not forget to regroup when necessary.

45 29 +57 = **131**	47 56 +83 = **186**	96 85 +69 = **250**	78 54 +73 = **205**	58 73 +69 = **200**
661 219 +452 = **1332**	465 352 +879 = **1696**	478 509 +382 = **1369**	823 756 +417 = **1996**	829 763 +127 = **1719**
589 473 +831 = **1893**	255 419 +799 = **1473**	387 876 +624 = **1887**	857 982 +156 = **1995**	229 802 +560 = **1591**

648 + 355 + 771 = **1774**   697 + 164 + 543 = **1404**

535 + 675 + 378 = **1588**   432 + 699 + 368 = **1499**

346 + 509 + 483 = **1338**   558 + 833 + 257 = **1648**

398 + 159 + 725 = **1282**   733 + 412 + 604 = **1749**

548 + 678 + 248 = **1474**   635 + 298 + 773 = **1706**

Name _____ Skill: Regrouping Subtraction Practice

RSP

Subtract the following problems. Do not forget to regroup when necessary.

457 -186 = **271**	809 -372 = **437**	515 -257 = **258**	300 -197 = **103**	921 -569 = **352**
324 -187 = **137**	698 -389 = **309**	914 -258 = **656**	736 -497 = **239**	815 -428 = **387**
802 -339 = **463**	774 -269 = **505**	512 -467 = **45**	439 -358 = **81**	705 -337 = **368**
603 -498 = **105**	881 -512 = **369**	840 -648 = **192**	637 -588 = **49**	724 -436 = **288**

929 - 464 = **465**   952 - 626 = **326**

808 - 749 = **59**   723 - 674 = **49**

630 - 252 = **378**   564 - 375 = **189**

700 - 411 = **289**   444 - 228 = **216**

830 - 541 = **289**   846 - 598 = **248**

Answer Key

Worksheet 5

Name_____ Skill: Regrouping Subtraction Practice

RSP
Subtract the following problems. Do not forget to regroup when necessary.

1,868 − 799 = 1069	2,452 − 1,875 = 577	8,025 − 3,562 = 4463	7,524 − 4,965 = 2559	9,342 − 4,786 = 4556
9,617 − 4,828 = 4789	3,806 − 1,497 = 2309	7,320 − 4,566 = 2754	9,064 − 2,278 = 6786	5,436 − 4,557 = 879
8,914 − 4,937 = 3977	5,088 − 1,329 = 3759	7,307 − 6,798 = 509	4,582 − 2,776 = 1806	9,520 − 3,745 = 5775
4,562 − 1,856 = 2706	6,407 − 3,819 = 2588	5,200 − 4,982 = 218	8,315 − 5,276 = 3039	9,028 − 4,346 = 4682

7,072 − 6,884 = 188   8,004 − 5,645 = 2359

9,426 − 2,558 = 6868   5,735 − 1,768 = 3967

6,421 − 3,757 = 2664   7,015 − 2,348 = 4667

6,624 − 1,999 = 4625   4,126 − 1,689 = 2437

7,327 − 5,698 = 1629   5,321 − 3,993 = 1328

©1996 Kelley Wingate Publications   5   KW 1304

Worksheet 6

Name_____ Skill: Regrouping Subtraction Practice

RSP
Subtract the following problems. Do not forget to regroup when necessary.

3,304 − 1,178 = 2126	5,100 − 2,487 = 2613	6,321 − 4,375 = 1946	7,034 − 3,158 = 3876	8,723 − 2,876 = 5847
9,014 − 3,658 = 5356	5,601 − 4,996 = 605	3,040 − 2,183 = 857	8,007 − 5,389 = 2618	5,432 − 2,848 = 2584
36,050 − 18,765 = 17,285	32,604 − 15,816 = 16,788	39,701 − 15,082 = 24,619	53,600 − 38,714 = 14,886	81,335 − 33,567 = 47,768
20,306 − 17,428 = 2878	80,040 − 69,367 = 10,673	73,148 − 42,349 = 30,799	90,004 − 73,665 = 16,339	41,025 − 28,357 = 12,668

5,134 − 1,458 = 3676   7,550 − 2,676 = 4874

8,044 − 3,769 = 4275   8,266 − 6,874 = 1392

20,603 − 16,799 = 3804   53,119 − 13,382 = 39,737

51,267 − 28,478 = 22,789   38,411 − 19,904 = 18,507

27,185 − 11,786 = 15,399   46,323 − 18,775 = 27,548

©1996 Kelley Wingate Publications   6   KW 1304

Worksheet 7

Name_____ Skill: Regrouping Subtraction Practice

RSP
Subtract the following problems. Do not forget to regroup when necessary.

3,080 − 1,982 = 1098	7,002 − 3,694 = 3308	5,401 − 4,378 = 1023	8,984 − 6,896 = 2086	5,004 − 2,786 = 2218
18,463 − 9,675 = 8788	47,031 − 39,554 = 7477	70,306 − 42,957 = 27,349	51,006 − 38,498 = 12,508	90,340 − 68,565 = 21,775
80,001 − 78,327 = 1674	40,102 − 23,487 = 16,615	73,004 − 59,775 = 13,229	40,300 − 29,472 = 10,828	80,404 − 55,652 = 24,752
320,432 − 179,385 = 141,047	263,143 − 177,418 = 85,725	631,227 − 580,669 = 50,558	554,654 − 165,988 = 388,666	723,072 − 395,184 = 327,888

7,432 − 5,554 = 1878   8,436 − 2,958 = 5478

48,702 − 13,825 = 34,877   84,251 − 26,744 = 57,507

70,345 − 42,557 = 27,788   93,126 − 57,849 = 35,277

153,424 − 133,667 = 19,757   325,471 − 199,684 = 125,787

237,143 − 181,678 = 55,465   416,314 − 278,745 = 137,569

©1996 Kelley Wingate Publications   7   KW 1304

Worksheet 8

Name_____ Skill: Regroup Add and Subtract Practice

RAP/RSP
Add or subtract the following problems. Do not forget to regroup when necessary.

701 − 482 = 219	684 + 398 = 1082	578 + 922 = 1500	900 − 187 = 713	452 − 378 = 74
5,015 + 3,276 = 8291	4,402 − 1,986 = 2416	5,917 + 5,687 = 11,604	8,003 − 2,576 = 5427	5,627 + 9,359 = 14,986
4,321 + 9,821 = 14,142	5,917 + 2,393 = 8310	7,403 − 6,776 = 627	5,715 + 3,892 = 9607	9,008 − 4,329 = 4679
7,092 − 2,296 = 4796	3,072 + 7,984 = 11,056	6,336 − 3,597 = 2739	4,624 + 8,835 = 13,459	9,723 − 4,738 = 4985

8,256 − 2,567 = 5689   5,884 + 3,726 = 2158

4,722 + 9,306 = 14,028   9,313 − 1,838 = 7475

7,112 − 6,207 = 905   2,659 + 7,482 = 10,141

6,345 − 2,857 = 3488   6,258 + 8,733 = 14,991

5,169 + 6,828 = 11,997   4,027 − 1,898 = 2129

©1996 Kelley Wingate Publications   8   KW 1304

CD-3724

Answer Key

Worksheet 1 (page 9)

Name _____ Skill: Regroup Add and Subtract Practice

RAP/RSP
Add or subtract the following problems. Do not forget to regroup when necessary.

702 - 554 **148**	327 + 492 **819**	840 - 338 **502**	761 - 286 **475**	497 + 615 **1112**
4,749 + 8,298 **13,047**	9,006 - 3,577 **5429**	7,668 + 3,942 **11,610**	8,254 - 6,268 **1986**	7,449 + 7,258 **14,707**
30,408 - 19,219 **11,189**	43,576 + 87,327 **130,903**	58,000 - 42,315 **15,685**	35,542 + 18,991 **54,533**	50,102 - 36,826 **13,276**
900,645 - 382,569 **518,076**	987,645 - 398,936 **588,709**	865,098 - 258,219 **606,879**	540,723 + 369,288 **910,011**	319,459 + 274,553 **594,012**

723 - 557 = **166**   38,965 + 72,558 = **111,523**

842 + 396 = **1238**   39,457 + 51,338 = **90,795**

5,332 + 1,768 = **7100**   72,436 - 27,597 = **44,839**

8,104 - 5,448 = **2656**   234,503 + 118,262 = **352,765**

81,406 - 59,818 = **21,588**   642,303 - 154,614 = **487,689**

©1996 Kelley Wingate Publications   9   KW 1304

Worksheet 2 (page 10)

Name _____ Skill: Regrouping Multiplication

RMP
How quickly can you complete this page? Time yourself. Ready, set, go!
Time : _____
Number Correct : _____

34 x 6 **204**	67 x 3 **201**	42 x 4 **168**	58 x 5 **290**	29 x 4 **116**	74 x 2 **148**	33 x 5 **165**	68 x 3 **204**
83 x 6 **498**	49 x 2 **98**	57 x 5 **285**	39 x 3 **117**	58 x 4 **232**	98 x 5 **490**	67 x 6 **402**	45 x 3 **135**
27 x 6 **162**	43 x 5 **215**	29 x 4 **116**	73 x 5 **365**	58 x 4 **232**	63 x 6 **378**	92 x 5 **460**	46 x 5 **230**
55 x 4 **220**	38 x 6 **228**	27 x 5 **135**	42 x 6 **252**	57 x 6 **342**	83 x 5 **415**	78 x 6 **468**	56 x 4 **224**

19 x 4 = **76**   48 x 5 = **240**   27 x 3 = **81**

67 x 6 = **402**   53 x 3 = **159**   54 x 4 = **216**

71 x 5 = **355**   63 x 6 = **378**   45 x 6 = **270**

37 x 5 = **185**   82 x 4 = **328**   29 x 3 = **87**

68 x 4 = **272**   43 x 3 = **129**   69 x 6 = **414**

©1996 Kelley Wingate Publications   10   KW 1304

Worksheet 3 (page 11)

Name _____ Skill: Regrouping Multiplication

RMP
How quickly can you complete this page? Time yourself. Ready, set, go!
Time : _____
Number Correct : _____

69 x 5 **345**	48 x 4 **192**	57 x 6 **342**	36 x 7 **252**	73 x 3 **219**	95 x 4 **380**	89 x 3 **267**	27 x 8 **216**
52 x 5 **260**	33 x 6 **198**	82 x 4 **328**	95 x 7 **665**	59 x 3 **177**	44 x 6 **264**	48 x 5 **240**	77 x 4 **308**
94 x 6 **564**	76 x 3 **228**	47 x 5 **235**	79 x 4 **316**	27 x 6 **162**	88 x 7 **616**	67 x 5 **335**	85 x 4 **340**
65 x 2 **130**	43 x 6 **258**	29 x 7 **203**	50 x 5 **250**	71 x 4 **284**	38 x 3 **152**	84 x 5 **420**	77 x 6 **462**

49 x 4 = **196**   27 x 5 = **135**   71 x 6 = **426**

39 x 3 = **117**   45 x 7 = **315**   54 x 2 = **108**

65 x 4 = **260**   71 x 3 = **213**   39 x 5 = **195**

83 x 4 = **332**   48 x 6 = **288**   63 x 7 = **441**

24 x 4 = **96**   65 x 3 = **195**   59 x 5 = **295**

©1996 Kelley Wingate Publications   11   KW 1304

Worksheet 4 (page 12)

Name _____ Skill: Regrouping Multiplication

RMP
How quickly can you complete this page? Time yourself. Ready, set, go!
Time : _____
Number Correct : _____

47 x 8 **376**	53 x 9 **477**	16 x 8 **128**	28 x 9 **252**	58 x 6 **348**	62 x 7 **434**	37 x 7 **259**	24 x 8 **192**
28 x 9 **252**	37 x 6 **222**	54 x 5 **270**	82 x 8 **656**	43 x 9 **387**	29 x 7 **203**	62 x 4 **248**	91 x 6 **546**
43 x 9 **387**	48 x 7 **336**	29 x 6 **174**	37 x 5 **185**	49 x 4 **245**	51 x 7 **357**	45 x 9 **405**	37 x 6 **222**
72 x 8 **576**	46 x 7 **322**	64 x 4 **256**	63 x 7 **441**	28 x 6 **168**	82 x 8 **656**	51 x 5 **255**	23 x 7 **161**

71 x 4 = **284**   37 x 9 = **333**   63 x 7 = **441**

45 x 6 = **270**   66 x 5 = **330**   84 x 6 = **504**

77 x 8 = **616**   59 x 7 = **413**   91 x 6 = **546**

56 x 4 = **224**   38 x 4 = **152**   48 x 5 = **240**

58 x 7 = **406**   78 x 9 = **702**   62 x 9 = **558**

©1996 Kelley Wingate Publications   12   KW 1304

CD-3724

Answer Key

Name _____
Skill: Regrouping Multiplication

RMP
How quickly can you complete this page? Time yourself. Ready, set, go!

Time : _____
Number Correct : _____

67 x 9	92 x 8	73 x 8	48 x 7	92 x 9	63 x 6	58 x 8	73 x 7
603	644	584	336	828	378	464	511

53 x 9	48 x 6	78 x 5	89 x 8	53 x 9	48 x 7	55 x 5	87 x 6
477	288	390	712	477	336	275	522

28 x 9	56 x 7	83 x 6	72 x 5	56 x 4	81 x 7	53 x 9	88 x 8
252	392	498	360	224	567	477	704

42 x 8	35 x 7	44 x 4	77 x 7	94 x 6	28 x 8	39 x 5	99 x 9
336	245	176	539	564	224	195	891

71 x 9 = 639 25 x 8 = 200 52 x 6 = 312
34 x 5 = 170 75 x 4 = 300 93 x 5 = 465
68 x 7 = 476 68 x 6 = 408 82 x 5 = 410
65 x 9 = 585 47 x 9 = 423 39 x 4 = 156
69 x 6 = 414 67 x 8 = 536 53 x 8 = 424

Name _____
Skill: Regrouping Multiplication

Multiply! Multiply!
Multiply the following problems. Do not forget to regroup when necessary.

40 x 38	59 x 15	84 x 35	76 x 38	90 x 67
1520	885	2940	2888	6030

24 x 86	96 x 33	72 x 64	85 x 17	21 x 39
863	3168	4608	1445	819

48 x 27	36 x 72	54 x 56	32 x 49	53 x 47
1296	2592	3024	1568	2491

80 x 98	36 x 52	29 x 16	32 x 44	53 x 45
7840	1872	464	408	2385

69 x 41 = 2829   37 x 46 = 1702
55 x 17 = 935   34 x 26 = 884
43 x 49 = 2107   16 x 88 = 1408
58 x 22 = 1276   37 x 42 = 1554
45 x 28 = 1260   56 x 25 = 1400

Name _____
Skill: Regrouping Multiplication

Multiply! Multiply!
Multiply the following problems. Do not forget to regroup when necessary.

37 x 12	15 x 20	43 x 25	61 x 38	44 x 94
444	300	1075	2318	4136

75 x 26	29 x 94	37 x 60	53 x 84	70 x 16
1950	2726	2220	4452	1120

19 x 67	98 x 32	54 x 53	74 x 59	67 x 38
1273	3136	2862	4366	2546

62 x 16	54 x 30	68 x 22	83 x 32	71 x 35
992	1620	1496	2656	2485

96 x 12 = 1152   48 x 25 = 1200
64 x 26 = 1664   45 x 15 = 675
54 x 38 = 2052   27 x 72 = 1944
47 x 31 = 1457   48 x 33 = 1584
56 x 17 = 952   64 x 36 = 2304

Name _____
Skill: Regrouping Multiplication

Multiply! Multiply!
Multiply the following problems. Do not forget to regroup when necessary.

405 x 32	315 x 74	923 x 85	337 x 56	560 x 62
12,960	23,310	78,455	18,872	34,720

617 x 75	479 x 24	982 x 37	189 x 45	303 x 41
46,275	11,496	36,334	8505	12,423

534 x 83	457 x 94	816 x 56	374 x 13	786 x 48
44,322	42,958	45,696	4862	37,728

295 x 64	315 x 32	447 x 25	379 x 14	386 x 42
18,880	10,080	11,175	5306	16,212

179 x 43 = 7697   317 x 52 = 16,484
825 x 19 = 15,675 364 x 27 = 9828
452 x 28 = 12,656 426 x 42 = 17,892
638 x 36 = 22,968 328 x 49 = 16,072
145 x 44 = 6380   276 x 32 = 8832

CD-3724

Answer Key

Name _____   Skill: Regrouping Multiplication

Multiply! Multiply!
Multiply the following problems. Do not forget to regroup when necessary.

615	658	396	447	827
x 37	x 26	x 42	x 67	x 43
22,755	17,108	16,632	29,949	35,561

513	365	412	307	724
x 74	x 29	x 48	x 75	x 59
37,962	10,585	19,776	23,025	42,716

382	391	188	507	129
x 86	x 42	x 92	x 68	x 38
32,852	16,422	17,296	38,025	4402

767	652	724	883	732
x 35	x 43	x 53	x 64	x 79
26,845	28,036	38,372	56,512	57,828

459 x 37 = 16,983   228 x 32 = 7296

946 x 58 = 54,868   673 x 28 = 18,844

179 x 36 = 6444   875 x 49 = 42,875

286 x 54 = 15,444   378 x 58 = 21,924

546 x 94 = 51,324   638 x 78 = 49,764

©1996 Kelley Wingate Publications   17   KW 1304

Name _____   Skill: Division - No Remainder

Dandy Division
Divide the following problems. Show your work.

$5\overline{)25}$ = 5	$7\overline{)28}$ = 4	$9\overline{)54}$ = 6	$6\overline{)48}$ = 8	$4\overline{)36}$ = 9
$8\overline{)40}$ = 5	$5\overline{)55}$ = 11	$8\overline{)64}$ = 8	$7\overline{)49}$ = 7	$9\overline{)18}$ = 2
$5\overline{)30}$ = 6	$12\overline{)60}$ = 5	$11\overline{)77}$ = 7	$4\overline{)48}$ = 12	$10\overline{)90}$ = 9
$6\overline{)60}$ = 10	$8\overline{)40}$ = 5	$4\overline{)36}$ = 9	$6\overline{)42}$ = 7	$3\overline{)36}$ = 12

63 ÷ 7 = 9   15 ÷ 3 = 5   55 ÷ 5 = 11

45 ÷ 9 = 5   48 ÷ 8 = 6   16 ÷ 4 = 4

12 ÷ 2 = 6   35 ÷ 7 = 5   56 ÷ 8 = 7

32 ÷ 4 = 8   99 ÷ 9 = 11   28 ÷ 7 = 4

20 ÷ 5 = 4   42 ÷ 6 = 7   18 ÷ 3 = 6

©1996 Kelley Wingate Publications   18   KW 1304

Name _____   Skill: Division - No Remainder

Dandy Division
Divide the following problems. Show your work.

$4\overline{)44}$ = 11	$9\overline{)99}$ = 11	$2\overline{)74}$ = 37	$6\overline{)90}$ = 15	$3\overline{)45}$ = 15
$7\overline{)70}$ = 10	$5\overline{)80}$ = 16	$3\overline{)57}$ = 19	$4\overline{)60}$ = 15	$9\overline{)90}$ = 10
$6\overline{)78}$ = 13	$8\overline{)80}$ = 10	$7\overline{)91}$ = 13	$6\overline{)60}$ = 10	$4\overline{)72}$ = 18
$3\overline{)39}$ = 13	$5\overline{)55}$ = 11	$4\overline{)76}$ = 19	$8\overline{)72}$ = 9	$9\overline{)63}$ = 7

42 ÷ 2 = 21   75 ÷ 3 = 25   56 ÷ 4 = 14

65 ÷ 5 = 13   72 ÷ 6 = 12   84 ÷ 7 = 12

88 ÷ 8 = 11   90 ÷ 9 = 10   96 ÷ 8 = 12

77 ÷ 7 = 11   84 ÷ 6 = 14   75 ÷ 5 = 15

48 ÷ 4 = 12   54 ÷ 3 = 18   54 ÷ 2 = 27

©1996 Kelley Wingate Publications   19   KW 1304

Name _____   Skill: Division - No Remainder

Dandy Division
Divide the following problems. Show your work.

$3\overline{)333}$ = 111	$5\overline{)305}$ = 61	$8\overline{)856}$ = 107	$7\overline{)497}$ = 71	$9\overline{)549}$ = 61
$5\overline{)455}$ = 91	$8\overline{)728}$ = 91	$6\overline{)648}$ = 108	$3\overline{)279}$ = 93	$2\overline{)216}$ = 108
$7\overline{)728}$ = 104	$9\overline{)189}$ = 21	$4\overline{)328}$ = 82	$3\overline{)912}$ = 304	$2\overline{)642}$ = 321

459 ÷ 9 = 51   219 ÷ 3 = 73

864 ÷ 8 = 108   628 ÷ 2 = 314

126 ÷ 6 = 21   714 ÷ 7 = 102

288 ÷ 4 = 72   168 ÷ 8 = 21

545 ÷ 5 = 109   426 ÷ 6 = 71

©1996 Kelley Wingate Publications   20   KW 1304

CD-3724

Answer Key

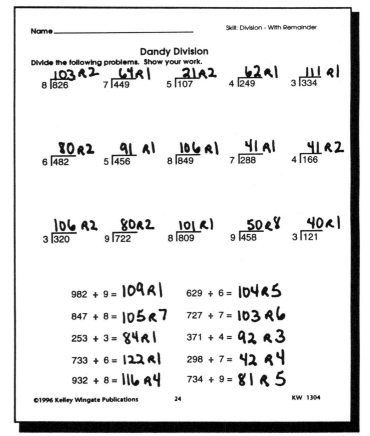

Worksheet (page 21)

Name _____   Skill: Division - No Remainder

Dandy Division
Divide the following problems. Show your work.

- $6\overline{)726} = 121$
- $8\overline{)256} = 32$
- $8\overline{)168} = 21$
- $8\overline{)968} = 121$
- $9\overline{)558} = 62$

- $7\overline{)854} = 122$
- $5\overline{)225} = 45$
- $7\overline{)364} = 52$
- $9\overline{)387} = 43$
- $3\overline{)552} = 184$

- $5\overline{)735} = 147$
- $4\overline{)576} = 144$
- $6\overline{)384} = 64$
- $3\overline{)852} = 284$
- $7\overline{)301} = 43$

$336 \div 8 = 42$ $224 \div 7 = 32$
$468 \div 9 = 52$ $628 \div 4 = 157$
$180 \div 4 = 45$ $252 \div 6 = 42$
$228 \div 3 = 76$ $192 \div 8 = 24$
$735 \div 5 = 147$ $258 \div 3 = 86$

©1996 Kelley Wingate Publications   21   KW 1304

Worksheet (page 22)

Name _____   Skill: Division - With Remainder

Dandy Division
Divide the following problems. Show your work.

- $3\overline{)28} = 9\ R1$
- $3\overline{)14} = 4\ R2$
- $6\overline{)51} = 8\ R3$
- $4\overline{)31} = 7\ R3$
- $2\overline{)17} = 8\ R1$

- $7\overline{)45} = 6\ R3$
- $4\overline{)25} = 6\ R1$
- $2\overline{)11} = 5\ R1$
- $8\overline{)47} = 5\ R7$
- $3\overline{)16} = 5\ R1$

- $6\overline{)20} = 3\ R2$
- $4\overline{)19} = 4\ R3$
- $5\overline{)33} = 6\ R3$
- $3\overline{)26} = 8\ R2$
- $7\overline{)50} = 7\ R1$

- $7\overline{)45} = 6\ R3$
- $2\overline{)17} = 8\ R1$
- $5\overline{)59} = 11\ R4$
- $7\overline{)60} = 8\ R4$
- $6\overline{)33} = 5\ R3$

$23 \div 2 = 11\ R1$ $10 \div 3 = 3\ R1$ $21 \div 4 = 5\ R1$
$48 \div 5 = 9\ R3$ $41 \div 6 = 6\ R5$ $37 \div 7 = 5\ R2$
$39 \div 8 = 4\ R7$ $50 \div 9 = 5\ R5$ $52 \div 8 = 6\ R4$
$45 \div 7 = 6\ R3$ $40 \div 6 = 6\ R4$ $33 \div 5 = 6\ R3$
$27 \div 4 = 6\ R3$ $14 \div 3 = 4\ R2$ $11 \div 2 = 5\ R1$

©1996 Kelley Wingate Publications   22   KW 1304

Worksheet (page 23)

Name _____   Skill: Division - With Remainder

Dandy Division
Divide the following problems. Show your work.

- $9\overline{)25} = 2\ R7$
- $6\overline{)39} = 6\ R3$
- $7\overline{)51} = 7\ R2$
- $9\overline{)61} = 6\ R7$
- $2\overline{)45} = 22\ R1$

- $4\overline{)54} = 13\ R2$
- $5\overline{)64} = 12\ R4$
- $8\overline{)89} = 11\ R1$
- $7\overline{)93} = 13\ R2$
- $3\overline{)29} = 9\ R2$

- $9\overline{)58} = 6\ R4$
- $8\overline{)93} = 11\ R5$
- $5\overline{)73} = 14\ R3$
- $4\overline{)67} = 16\ R3$
- $7\overline{)36} = 5\ R1$

- $7\overline{)85} = 12\ R1$
- $9\overline{)80} = 8\ R8$
- $3\overline{)64} = 21\ R1$
- $9\overline{)84} = 9\ R3$
- $6\overline{)82} = 13\ R4$

$63 \div 2 = 31\ R1$ $98 \div 9 = 10\ R8$ $81 \div 7 = 11\ R4$
$53 \div 4 = 13\ R1$ $57 \div 2 = 28\ R1$ $73 \div 4 = 18\ R1$
$96 \div 7 = 13\ R5$ $89 \div 5 = 17\ R4$ $65 \div 8 = 8\ R1$
$67 \div 3 = 22\ R1$ $59 \div 8 = 7\ R3$ $75 \div 9 = 8\ R3$
$93 \div 6 = 15\ R3$ $76 \div 6 = 12\ R4$ $93 \div 2 = 46\ R1$

©1996 Kelley Wingate Publications   23   KW 1304

Worksheet (page 24)

Name _____   Skill: Division - With Remainder

Dandy Division
Divide the following problems. Show your work.

- $8\overline{)826} = 103\ R2$
- $7\overline{)449} = 64\ R1$
- $5\overline{)107} = 21\ R2$
- $4\overline{)249} = 62\ R1$
- $3\overline{)334} = 111\ R1$

- $6\overline{)482} = 80\ R2$
- $5\overline{)456} = 91\ R1$
- $8\overline{)849} = 106\ R1$
- $7\overline{)288} = 41\ R1$
- $4\overline{)166} = 41\ R2$

- $3\overline{)320} = 106\ R2$
- $9\overline{)722} = 80\ R2$
- $8\overline{)809} = 101\ R1$
- $9\overline{)458} = 50\ R8$
- $3\overline{)121} = 40\ R1$

$982 \div 9 = 109\ R1$ $629 \div 6 = 104\ R5$
$847 \div 8 = 105\ R7$ $727 \div 7 = 103\ R6$
$253 \div 3 = 84\ R1$ $371 \div 4 = 92\ R3$
$733 \div 6 = 122\ R1$ $298 \div 7 = 42\ R4$
$932 \div 8 = 116\ R4$ $734 \div 9 = 81\ R5$

©1996 Kelley Wingate Publications   24   KW 1304

Answer Key

Dandy Division
Divide the following problems. Show your work.

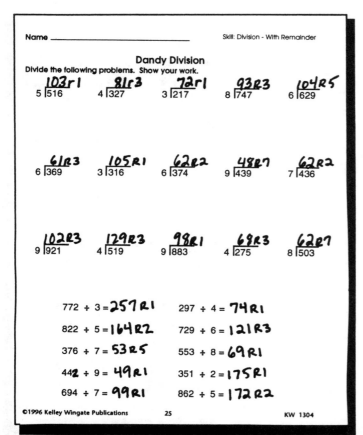

103r1	81r3	72r1	93R3	104R5
5⟌516	4⟌327	3⟌217	8⟌747	6⟌629
61R3	105R1	62R2	48R7	62R2
6⟌369	3⟌316	6⟌374	9⟌439	7⟌436
102R3	129R3	98R1	69R3	62R7
9⟌921	4⟌519	9⟌883	4⟌275	8⟌503

772 ÷ 3 = **257R1**   297 ÷ 4 = **74R1**

822 ÷ 5 = **164R2**   729 ÷ 6 = **121R3**

376 ÷ 7 = **53R5**   553 ÷ 8 = **69R1**

442 ÷ 9 = **49R1**   351 ÷ 2 = **175R1**

694 ÷ 7 = **99R1**   862 ÷ 5 = **172R2**

Hopscotch
Use your math facts to complete the hopscotch board.

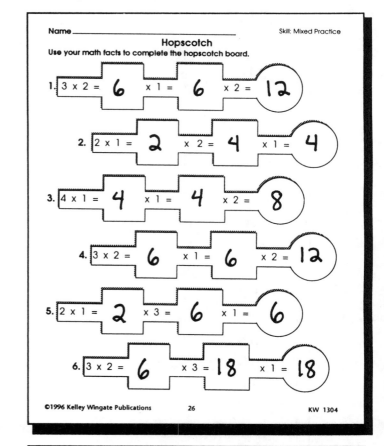

1. 3 x 2 = **6**   x 1 = **6**   x 2 = **12**
2. 2 x 1 = **2**   x 2 = **4**   x 1 = **4**
3. 4 x 1 = **4**   x 1 = **4**   x 2 = **8**
4. 3 x 2 = **6**   x 1 = **6**   x 2 = **12**
5. 2 x 1 = **2**   x 3 = **6**   x 1 = **6**
6. 3 x 2 = **6**   x 3 = **18**   x 1 = **18**

Hopscotch
Use your math facts to complete the hopscotch board.

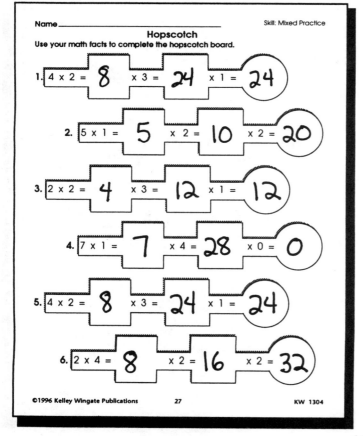

1. 4 x 2 = **8**   x 3 = **24**   x 1 = **24**
2. 5 x 1 = **5**   x 2 = **10**   x 2 = **20**
3. 2 x 2 = **4**   x 3 = **12**   x 1 = **12**
4. 7 x 1 = **7**   x 4 = **28**   x 0 = **0**
5. 4 x 2 = **8**   x 3 = **24**   x 1 = **24**
6. 2 x 4 = **8**   x 2 = **16**   x 2 = **32**

Hopscotch
Use your math facts to complete the hopscotch board.

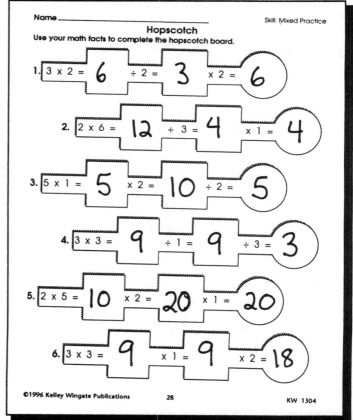

1. 3 x 2 = **6**   ÷ 2 = **3**   x 2 = **6**
2. 2 x 6 = **12**   ÷ 3 = **4**   x 1 = **4**
3. 5 x 1 = **5**   x 2 = **10**   ÷ 2 = **5**
4. 3 x 3 = **9**   ÷ 1 = **9**   ÷ 3 = **3**
5. 2 x 5 = **10**   x 2 = **20**   x 1 = **20**
6. 3 x 3 = **9**   x 1 = **9**   x 2 = **18**

Answer Key

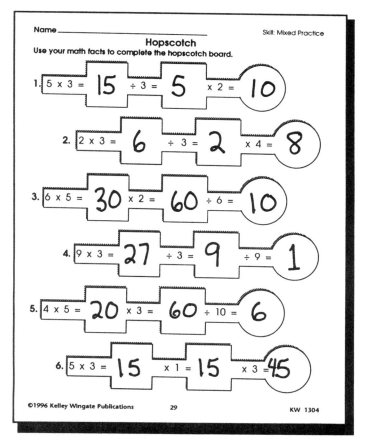

Name _____

Hopscotch

Skill: Mixed Practice

Use your math facts to complete the hopscotch board.

1. 5 x 3 = **15** ÷ 3 = **5** x 2 = **10**
2. 2 x 3 = **6** ÷ 3 = **2** x 4 = **8**
3. 6 x 5 = **30** x 2 = **60** ÷ 6 = **10**
4. 9 x 3 = **27** ÷ 3 = **9** ÷ 9 = **1**
5. 4 x 5 = **20** x 3 = **60** ÷ 10 = **6**
6. 5 x 3 = **15** x 1 = **15** x 3 = **45**

©1996 Kelley Wingate Publications   29   KW 1304

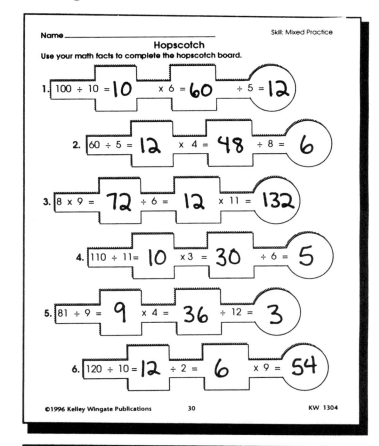

Name _____

Hopscotch

Skill: Mixed Practice

Use your math facts to complete the hopscotch board.

1. 100 ÷ 10 = **10** x 6 = **60** ÷ 5 = **12**
2. 60 ÷ 5 = **12** x 4 = **48** ÷ 8 = **6**
3. 8 x 9 = **72** ÷ 6 = **12** x 11 = **132**
4. 110 ÷ 11 = **10** x 3 = **30** ÷ 6 = **5**
5. 81 ÷ 9 = **9** x 4 = **36** ÷ 12 = **3**
6. 120 ÷ 10 = **12** ÷ 2 = **6** x 9 = **54**

©1996 Kelley Wingate Publications   30   KW 1304

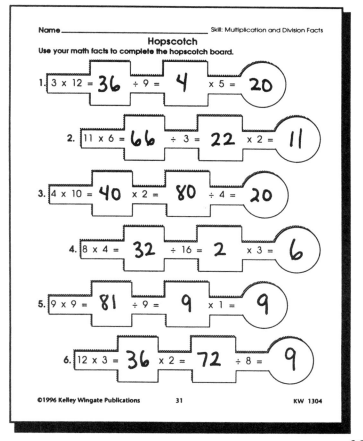

Name _____

Skill: Multiplication and Division Facts

Hopscotch

Use your math facts to complete the hopscotch board.

1. 3 x 12 = **36** ÷ 9 = **4** x 5 = **20**
2. 11 x 6 = **66** ÷ 3 = **22** x 2 = **11**
3. 4 x 10 = **40** x 2 = **80** ÷ 4 = **20**
4. 8 x 4 = **32** ÷ 16 = **2** x 3 = **6**
5. 9 x 9 = **81** ÷ 9 = **9** x 1 = **9**
6. 12 x 3 = **36** x 2 = **72** ÷ 8 = **9**

©1996 Kelley Wingate Publications   31   KW 1304

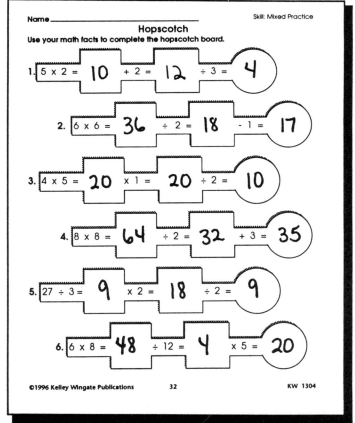

Name _____

Skill: Mixed Practice

Hopscotch

Use your math facts to complete the hopscotch board.

1. 5 x 2 = **10** + 2 = **12** ÷ 3 = **4**
2. 6 x 6 = **36** ÷ 2 = **18** - 1 = **17**
3. 4 x 5 = **20** x 1 = **20** ÷ 2 = **10**
4. 8 x 8 = **64** ÷ 2 = **32** + 3 = **35**
5. 27 ÷ 3 = **9** x 2 = **18** ÷ 2 = **9**
6. 6 x 8 = **48** ÷ 12 = **4** x 5 = **20**

©1996 Kelley Wingate Publications   32   KW 1304

CD-3724

Answer Key

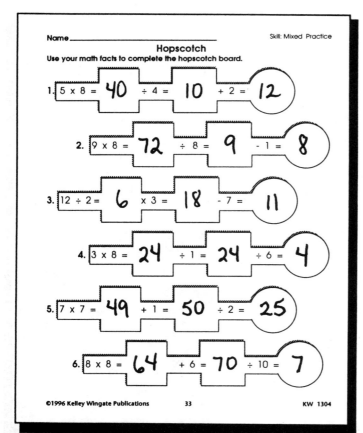

Name _____ Skill: Mixed Practice
Hopscotch
Use your math facts to complete the hopscotch board.

1. $5 \times 8 = $ **40** $\div 4 = $ **10** $+ 2 = $ **12**

2. $9 \times 8 = $ **72** $\div 8 = $ **9** $- 1 = $ **8**

3. $12 \div 2 = $ **6** $\times 3 = $ **18** $- 7 = $ **11**

4. $3 \times 8 = $ **24** $\div 1 = $ **24** $\div 6 = $ **4**

5. $7 \times 7 = $ **49** $+ 1 = $ **50** $\div 2 = $ **25**

6. $8 \times 8 = $ **64** $+ 6 = $ **70** $\div 10 = $ **7**

©1996 Kelley Wingate Publications   33   KW 1304

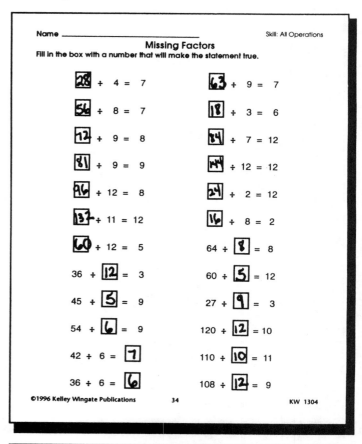

Name _____ Skill: All Operations
Missing Factors
Fill in the box with a number that will make the statement true.

28 $\div 4 = 7$	**63** $\div 9 = 7$
56 $\div 8 = 7$	**18** $\div 3 = 6$
72 $\div 9 = 8$	**84** $\div 7 = 12$
81 $\div 9 = 9$	**144** $\div 12 = 12$
96 $\div 12 = 8$	**24** $\div 2 = 12$
132 $\div 11 = 12$	**16** $\div 8 = 2$
60 $\div 12 = 5$	$64 \div$ **8** $= 8$
$36 \div$ **12** $= 3$	$60 \div$ **5** $= 12$
$45 \div$ **5** $= 9$	$27 \div$ **9** $= 3$
$54 \div$ **6** $= 9$	$120 \div$ **12** $= 10$
$42 \div 6 = $ **7**	$110 \div$ **10** $= 11$
$36 \div 6 = $ **6**	$108 \div$ **12** $= 9$

©1996 Kelley Wingate Publications   34   KW 1304

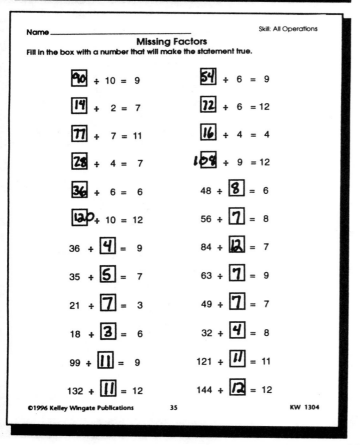

Name _____ Skill: All Operations
Missing Factors
Fill in the box with a number that will make the statement true.

90 $\div 10 = 9$	**54** $\div 6 = 9$
14 $\div 2 = 7$	**72** $\div 6 = 12$
77 $\div 7 = 11$	**16** $\div 4 = 4$
28 $\div 4 = 7$	**108** $\div 9 = 12$
36 $\div 6 = 6$	$48 \div$ **8** $= 6$
120 $\div 10 = 12$	$56 \div$ **7** $= 8$
$36 \div$ **4** $= 9$	$84 \div$ **12** $= 7$
$35 \div$ **5** $= 7$	$63 \div$ **7** $= 9$
$21 \div$ **7** $= 3$	$49 \div$ **7** $= 7$
$18 \div$ **3** $= 6$	$32 \div$ **4** $= 8$
$99 \div$ **11** $= 9$	$121 \div$ **11** $= 11$
$132 \div$ **11** $= 12$	$144 \div$ **12** $= 12$

©1996 Kelley Wingate Publications   35   KW 1304

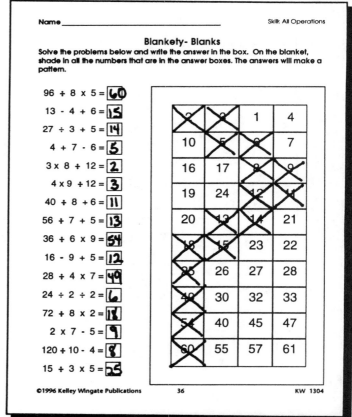

Name _____ Skill: All Operations
Blankety- Blanks
Solve the problems below and write the answer in the box. On the blanket, shade in all the numbers that are in the answer boxes. The answers will make a pattern.

$96 \div 8 \times 5 = $ **60**
$13 - 4 + 6 = $ **15**
$27 \div 3 + 5 = $ **14**
$4 + 7 - 6 = $ **5**
$3 \times 8 \div 12 = $ **2**
$4 \times 9 \div 12 = $ **3**
$40 \div 8 + 6 = $ **11**
$56 \div 7 + 5 = $ **13**
$36 \div 6 \times 9 = $ **54**
$16 - 9 + 5 = $ **12**
$28 + 4 \times 7 = $ **49**
$24 \div 2 \div 2 = $ **6**
$72 \div 8 \times 2 = $ **18**
$2 \times 7 - 5 = $ **9**
$120 \div 10 - 4 = $ **8**
$15 \div 3 \times 5 = $ **25**

2	3	1	4
10	5	6	7
16	17	8	9
19	24	12	11
20	13	14	21
18	15	23	22
25	26	27	28
49	30	32	33
54	40	45	47
60	55	57	61

©1996 Kelley Wingate Publications   36   KW 1304

CD-3724

Answer Key

Name_____   Skill: All Operations

Blankety- Blanks

Solve the problems below and write the answer in the box. On the blanket,
shade in all the numbers that are in the answer boxes. The answers will make a
pattern.

Name_____   Skill: All Operations

Blankety- Blanks

Solve the problems below and write the answer in the box. On the blanket,
shade in all the numbers that are in the answer boxes. The answers will make a
pattern.

$4 \times 5 \div 10 =$ 2
$32 \div 8 \times 4 =$ 16
$2 \times 4 \times 8 =$ 64
$5 \times 8 \div 4 =$ 10
$77 \div 11 + 6 =$ 13
$96 \div 8 - 7 =$ 5
$8 \times 3 \div 4 =$ 6
$63 \div 9 + 8 =$ 15
$55 \div 5 - 4 =$ 7
$144 \div 12 \times 9 =$ 108
$14 - 6 + 3 =$ 11
$2 \times 3 \times 8 =$ 48
$7 \times 2 - 5 =$ 9
$36 \div 9 \times 3 =$ 12
$5 + 3 + 9 =$ 17
$132 \div 11 - 4 =$ 8

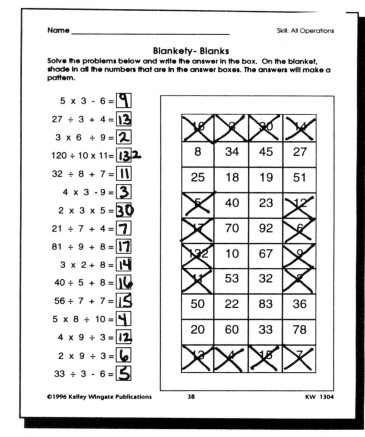

©1996 Kelley Wingate Publications   37   KW 1304

$5 \times 3 - 6 =$ 9
$27 \div 3 + 4 =$ 13
$3 \times 6 \div 9 =$ 2
$120 \div 10 \times 11 =$ 132
$32 \div 8 + 7 =$ 11
$4 \times 3 - 9 =$ 3
$2 \times 3 \times 5 =$ 30
$21 \div 7 + 4 =$ 7
$81 \div 9 + 8 =$ 17
$3 \times 2 + 8 =$ 14
$40 \div 5 + 8 =$ 16
$56 \div 7 + 7 =$ 15
$5 \times 8 \div 10 =$ 4
$4 \times 9 \div 3 =$ 12
$2 \times 9 \div 3 =$ 6
$33 \div 3 - 6 =$ 5

©1996 Kelley Wingate Publications   38   KW 1304

Name_____   Skill: All Operations

Compare Squares

Compare the number sentences. Write $<$, $>$, or $=$ in the square to make a
true math statement. The first problem is done for you.

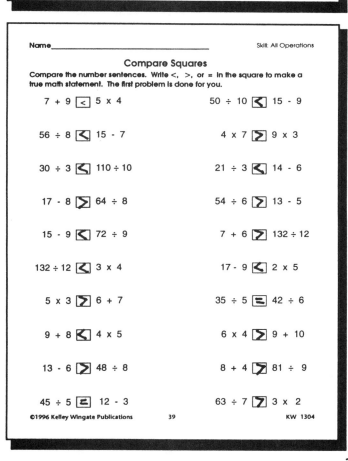

©1996 Kelley Wingate Publications   39   KW 1304

Name_____   Skill: All Operations

Compare Squares

Compare the number sentences. Write $<$, $>$, or $=$ in the square to make a
true math statement. The first problem is done for you.

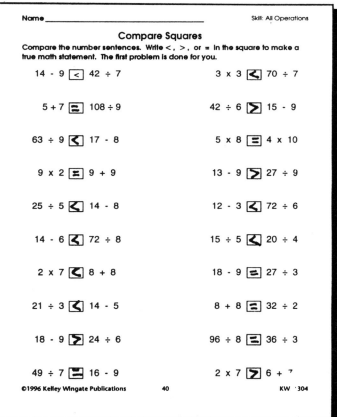

©1996 Kelley Wingate Publications   40   KW 1304

Answer Key

Name_____ Skill: All Operations

Compare Squares

Compare the number sentences. Write <, >, or = in the square to make a true math statement. The first problem is done for you.

13 - 7 [<] 49 ÷ 7   33 ÷ 11 [<] 3 x 3

21 ÷ 3 [>] 55 ÷ 11   12 ÷ 2 [<] 18 - 9

16 ÷ 4 [<] 2 x 3   2 x 9 [>] 7 + 9

9 + 9 [=] 3 x 6   9 + 2 [=] 44 ÷ 4

15 - 7 [>] 14 - 8   12 - 8 [>] 36 ÷ 12

14 - 9 [>] 20 ÷ 5   36 ÷ 6 [<] 4 x 3

3 x 8 [<] 5 x 5   5 x 6 [>] 7 x 4

96 ÷ 8 [>] 13 - 4   64 ÷ 8 [<] 32 ÷ 4

12 - 9 [>] 24 ÷ 12   16 - 9 [>] 45 ÷ 9

121 ÷ 11 [>] 16 - 8   8 x 6 [>] 16 - 7

Name_____ Skill: All Operations

Compare Squares

Compare the number sentences. Write <, >, or = in the square to make a true math statement. The first problem is done for you.

7 - 3 [<] 42 ÷ 7   16 - 7 [>] 72 ÷ 9

6 x 6 [>] 108 ÷ 9   110 ÷ 11 [>] 14 - 8

5 x 3 [>] 17 - 8   7 + 8 [<] 3 x 6

81 ÷ 9 [<] 9 + 9   17 - 9 [>] 42 ÷ 7

35 ÷ 7 [<] 14 - 8   9 x 4 [=] 6 x 6

14 - 8 [<] 72 ÷ 8   48 ÷ 8 [>] 24 ÷ 8

144 ÷ 12 [<] 8 + 8   13 - 9 [>] 21 ÷ 7

12 ÷ 3 [<] 14 - 5   6 x 3 [=] 9 + 9

7 + 4 [>] 24 ÷ 6   12 - 7 [<] 45 ÷ 5

16 - 9 [=] 14 - 7   4 x 3 [<] 8 + 5

Name_____ Skill: All Operations

Mystery Math

Look at the mystery number. Circle all math expressions in that row which equal the mystery number. The first problem is done for you.

Mystery Number	Math Expression			
3	28 ÷ 7	12 - 8	(27 ÷ 9)	(12 - 9)
11	(4 + 7)	(77 ÷ 7)	8 + 5	132 ÷ 12
4	15 - 9	(32 ÷ 8)	(12 - 8)	36 ÷ 6
6	15 - 8	24 ÷ 6	(3 x 2)	(14 - 8)
48	(4 x 12)	9 x 4	(8 x 6)	4 x 8
9	8 + 2	14 - 6	(63 ÷ 7)	(17 - 8)
12	(96 ÷ 9)	5 + 9	8 + 6	(4 x 3)
7	(16 - 9)	35 ÷ 7	(49 ÷ 7)	5 + 3
24	6 x 3	(2 x 12)	8 x 4	(4 x 6)

Name_____ Skill: All Operations

Mystery Math

Look at the mystery number. Circle all math expressions in that row which equal the mystery number. The first problem is done for you.

Mystery Number	Math Expression			
4	16 - 9	(12 - 8)	11 - 6	(16 ÷ 4)
11	5 + 7	(6 + 5)	(121 ÷ 11)	(4 + 7)
8	15 - 8	108 ÷ 12	(17 - 9)	(96 ÷ 12)
10	30 ÷ 5	(7 + 3)	(110 ÷ 11)	5 + 4
3	24 ÷ 6	(11 - 8)	(36 ÷ 12)	10 - 9
6	36 ÷ 9	(15 - 9)	(72 ÷ 12)	14 - 7
7	48 ÷ 6	(13 - 6)	(16 - 9)	42 ÷ 7
12	7 + 6	132 ÷ 12	(60 ÷ 5)	(120 ÷ 10)
5	(35 ÷ 7)	(14 - 9)	15 ÷ 5	30 ÷ 5

Answer Key

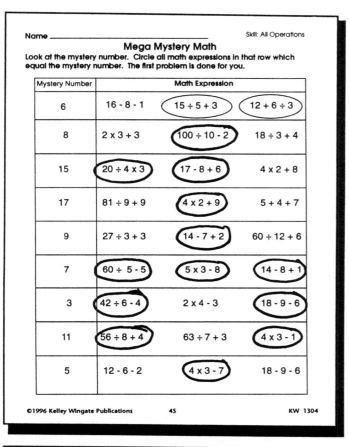

Name _____   Skill: All Operations

Mega Mystery Math

Look at the mystery number. Circle all math expressions in that row which equal the mystery number. The first problem is done for you.

Mystery Number	Math Expression		
6	16 - 8 - 1	(15 ÷ 5 + 3)	(12 + 6 ÷ 3)
8	2 x 3 + 3	(100 ÷ 10 - 2)	18 ÷ 3 + 4
15	(20 ÷ 4 x 3)	(17 - 8 + 6)	4 x 2 + 8
17	81 ÷ 9 + 9	(4 x 2 + 9)	5 + 4 + 7
9	27 ÷ 3 + 3	(14 - 7 + 2)	60 ÷ 12 + 6
7	(60 ÷ 5 - 5)	(5 x 3 - 8)	(14 - 8 + 1)
3	(42 ÷ 6 - 4)	2 x 4 - 3	(18 - 9 - 6)
11	(56 ÷ 8 + 4)	63 ÷ 7 + 3	(4 x 3 - 1)
5	12 - 6 - 2	(4 x 3 - 7)	18 - 9 - 6

©1996 Kelley Wingate Publications   45   KW 1304

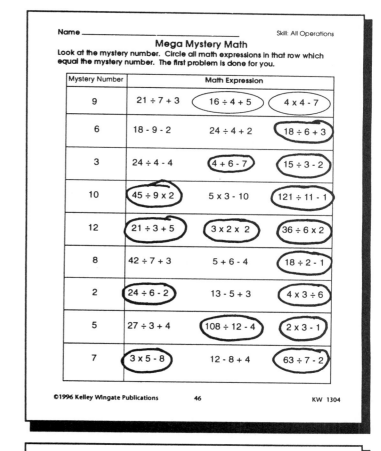

Name _____   Skill: All Operations

Mega Mystery Math

Look at the mystery number. Circle all math expressions in that row which equal the mystery number. The first problem is done for you.

Mystery Number	Math Expression		
9	21 ÷ 7 + 3	(16 ÷ 4 + 5)	(4 x 4 - 7)
6	18 - 9 - 2	24 ÷ 4 + 2	(18 ÷ 6 + 3)
3	24 ÷ 4 - 4	(4 + 6 - 7)	(15 ÷ 3 - 2)
10	(45 ÷ 9 x 2)	5 x 3 - 10	(121 ÷ 11 - 1)
12	(21 ÷ 3 + 5)	(3 x 2 x 2)	(36 ÷ 6 x 2)
8	42 ÷ 7 + 3	5 + 6 - 4	(18 ÷ 2 - 1)
2	(24 ÷ 6 - 2)	13 - 5 + 3	(4 x 3 ÷ 6)
5	27 ÷ 3 + 4	(108 ÷ 12 - 4)	(2 x 3 - 1)
7	(3 x 5 - 8)	12 - 8 + 4	(63 ÷ 7 - 2)

©1996 Kelley Wingate Publications   46   KW 1304

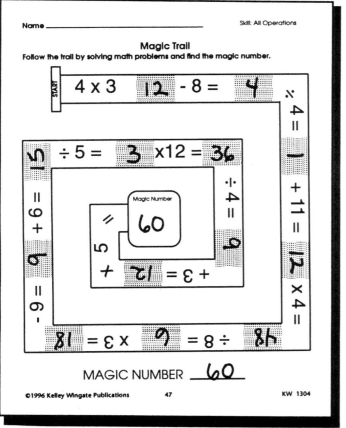

Name _____   Skill: All Operations

Magic Trail

Follow the trail by solving math problems and find the magic number.

START 4 x 3 **12** - 8 = **4**

MAGIC NUMBER __60__

©1996 Kelley Wingate Publications   47   KW 1304

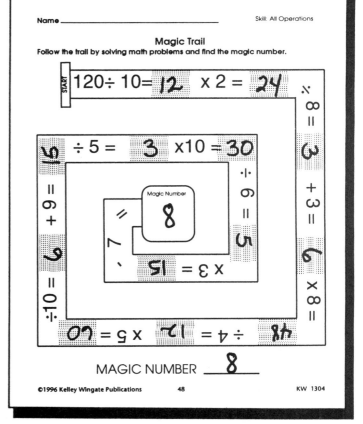

Name _____   Skill: All Operations

Magic Trail

Follow the trail by solving math problems and find the magic number.

START 120 ÷ 10 = **12** x 2 = **24**

MAGIC NUMBER __8__

©1996 Kelley Wingate Publications   48   KW 1304

©1996 Kelley Wingate Publications   CD-3724

Answer Key

Page 49

Name_____ Skill: Place Value

Place Space

| Millions | , | Hundred Thousands | Ten Thousands | Thousands | , | Hundreds | Tens | Ones | . | Tenths | Hundredths | Thousandths |

1. The number is: `7,320,196.485`

A. Name the digit in the tens place __9__
B. Name the digit in the tenths place __4__
C. Name the digit in the millions place __7__
D. Name the digit in the ones place __6__
E. In what place value is the digit "0"? __thousands__
F. In what place value is the digit "4"? __tenths__
G. In what place value is the digit "3"? __hundred thousands__
H. In what place value is the digit "5"? __thousandths__

2. The number is: `8,657.321`

A. Name the digit in the hundreds place __6__
B. Name the digit in the hundredths place __2__
C. Name the digit in the thousands place __8__
D. Name the digit in the tenths place __3__
E. Name the number that is one hundred more __8,757.321__
F. Name the number that is one thousand less __7,657.321__
G. Name the number that is one hundredth less __8,657.311__
H. Name the number that is one more __8,658.321__

3. The number is: `7,320,196.485`

A. Name the digit in the millions place __7__
B. Name the digit in the ones place __6__
C. Name the digit in the thousandths place __5__
D. Name the digit in the ten thousands place __2__
E. Name the number that is ten thousand less __7,310,196.485__
F. Name the number that is one thousandth more __7,320,196.486__
G. Name the number that is one less __7,320,195.485__
H. Name the number that is one million more __8,320,196.485__

©1996 Kelley Wingate Publications 49 KW 1304

Page 50

Name_____ Skill: Place Value

Place Space

| Millions | , | Hundred Thousands | Ten Thousands | Thousands | , | Hundreds | Tens | Ones | . | Tenths | Hundredths | Thousandths |

1. The number is: `3,649,158.072`

A. Name the digit in the hundred thousands place __6__
B. Name the digit in the one thousands place __9__
C. Name the digit in the tens place __5__
D. Name the digit in the hundredths place __7__
E. In what place value is the digit "0"? __tenths__
F. In what place value is the digit "4"? __ten thousand__
G. In what place value is the digit "3"? __millions__
H. In what place value is the digit "2"? __thousandths__

2. The number is: `17,348.620`

A. Name the digit in the hundreds place __3__
B. Name the digit in the thousandths place __0__
C. Name the digit in the ten thousands place __1__
D. Name the digit in the tenths place __6__
E. Name the number that is one thousand more __18,348.620__
F. Name the number that is one tenth less __17,348.520__
G. Name the number that is ten more __17,358.620__
H. Name the number that is one hundred less __17,248.620__

3. The number is: `9,806,432.517`

A. Name the digit in the hundred thousands place __8__
B. Name the digit in the ones place __2__
C. Name the digit in the millions place __9__
D. Name the digit in the thousandths place __7__
E. Name the number that is ten thousand more __9,816,432.517__
F. Name the number that is one thousand less __9,805,432.517__
G. Name the number that is one hundred less __9,806,332.517__
H. Name the number that is one million less __8,806,432.517__

©1996 Kelley Wingate Publications 50 KW 1304

Page 51

Name_____ Skill: Rounding Numbers

Rounding Round-Up

To round any number, follow these simple rules:

Underline the place value you are rounding to.
Circle the digit to the right of the underlined digit.
If the circled number is 0, 1, 2, 3, or 4 the underlined digit stays the same.
If the circled number is 5, 6, 7, 8, or 9 the underlined digit goes up by 1.
The circled digit and all digits to the right become a zero.

Round to the nearest ten:
364   366
364   366
36④   36⑥
360   370

1. Round these numbers to the nearest thousand:

A. 96,299 __96,000__   B. 34,941 __35,000__
C. 142,298 __142,000__   D. 3,850 __4,000__
E. 88,880 __89,000__   F. 9,551 __10,000__
G. 7,492 __7,000__   H. 32,621 __33,000__
I. 11,804 __12,000__   J. 27,700 __28,000__

2. Round these numbers to the nearest ten thousand:

A. 452,398 __450,000__   B. 986,334 __990,000__
C. 72,994 __70,000__   D. 296,303 __300,000__
E. 805,125 __810,000__   F. 546,498 __550,000__
G. 15,300 __20,000__   H. 1,832,759 __1,830,000__
I. 301,399 __300,000__   J. 3,488,621 __3,490,000__

3. Round these numbers to the nearest hundred thousand:

A. 43,645,117 __43,600,000__   B. 416,667 __400,000__
C. 6,285,136 __6,300,000__   D. 3,821,501 __3,800,000__
E. 19,915,324 __19,900,000__   F. 2,137,148 __2,100,000__
G. 743,698 __700,000__   H. 292,531 __300,000__
I. 461,433 __500,000__   J. 3,554,712 __3,600,000__

©1996 Kelley Wingate Publications 51 KW 1304

Page 52

Name_____ Skill: Rounding Numbers

Rounding Round-Up

To round any number, follow these simple rules:

Underline the place value you are rounding to.
Circle the digit to the right of the underlined digit.
If the circled number is 0, 1, 2, 3, or 4 the underlined digit stays the same.
If the circled number is 5, 6, 7, 8, or 9 the underlined digit goes up by 1.
The circled digit and all digits to the right become a zero.

Round to the nearest ten:
364   366
364   366
36④   36⑥
360   370

1. Round these numbers to the nearest tenth:

A. 56.324 __56.3__   B. 41.12 __41.1__
C. 18.76 __18.8__   D. 307.685 __307.7__
E. 4.945 __4.9__   F. 10.852 __10.9__
G. 132.75 __132.8__   H. 5.398 __5.4__
I. 60.36 __60.4__   J. 72.72 __72.7__

2. Round these numbers to the nearest hundredth:

A. 230.036 __230.04__   B. 155.872 __155.87__
C. 35.512 __35.51__   D. 1,725.977 __1,725.98__
E. 127.384 __127.38__   F. .296 __.30__
G. 59.305 __59.31__   H. 158.335 __158.34__
I. 82.361 __82.36__   J. 12.765 __12.77__

3. Round these numbers to the nearest whole number (one):

A. 15.356 __15__   B. 57.11 __57__
C. 37.84 __38__   D. 18.9 __19__
E. 109.288 __109__   F. 3.646 __4__
G. 722.510 __723__   H. 623.923 __624__
I. 78.41 __78__   J. 1,204.2 __1,204__

©1996 Kelley Wingate Publications 52 KW 1304

115

CD-3724

Answer Key

Worksheet 53

Name _____   Skill: Rounding Numbers

Rounding Round-Up

To round any number, follow these simple rules:

Underline the place value you are rounding to.
Circle the digit to the right of the underlined digit.
If the circled number is 0, 1, 2, 3, or 4 the underlined digit stays the same.
If the circled number is 5, 6, 7, 8, or 9 the underlined digit goes up by 1.
The circled digit and all digits to the right become a zero.

Round to the nearest ten:	
364	366
364	366
364	366
360	370

1. Round these numbers to the nearest thousandth:

A. 7.30452 __7.305__   B. 12.00453 __12.005__
C. 0.0151 __0.015__   D. 7.98221 __7.982__
E. 14.30649 __14.306__   F. 27.01056 __27.011__
G. 2.5428 __2.543__   H. 124.5793 __124.579__
I. 9.1342 __9.134__   J. 36.4468 __36.447__

2. Round these numbers to the nearest thousand:

A. 452,398 __452,000__   B. 986,334 __986,000__
C. 72,994 __73,000__   D. 296,303 __296,000__
E. 805,125 __805,000__   F. 546,498 __546,000__
G. 15,300 __15,000__   H. 1,832,759 __1,833,000__
I. 301,399 __301,000__   J. 3,488,621 __3,489,000__

3. Round these numbers to the nearest hundred thousand:

A. 624,677 __600,000__   B. 552,542 __600,000__
C. 285,453 __300,000__   D. 1,712,899 __1,700,000__
E. 681,398 __700,000__   F. 9,107,600 __9,100,000__
G. 519,601 __500,000__   H. 243,377 __200,000__
I. 154,429 __200,000__   J. 3,094,192 __3,100,000__

©1996 Kelley Wingate Publications   53   KW 1304

Worksheet 54

Name _____   Skill: Rounding and Estimating Numbers

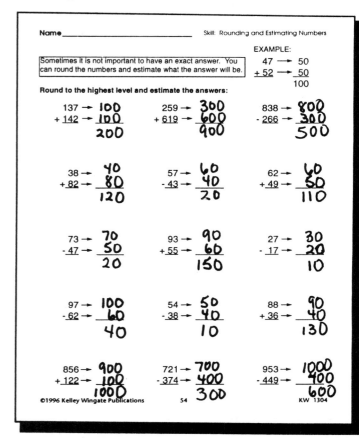

©1996 Kelley Wingate Publications   54   KW 1304

Worksheet 55

Name _____   Skill: Rounding and Estimating Numbers

Sometimes it is not important to have an exact answer. You can round the numbers and estimate what the answer will be.

EXAMPLE:
47 → 50
x 52 → 50
2500

Round to the highest level and estimate the answers:

629 → 600   718 → 700   608 → 600
+ 384 → 400   + 949 → 900   + 284 → 300
1000   1600   900

824 → 800   454 → 500   733 → 100
- 396 → 400   - 243 → 200   - 187 → 200
400   300   500

54 → 50   26 → 30   49 → 50
x 37 → 40   x 16 → 20   x 23 → 20
2000   600   1000

321 → 300   183 → 200   538 → 500
x 416 → 400   x 652 → 700   x 923 → 900
120,000   140,000   450,000

816 → 800   197 → 200   272 → 300
x 483 → 500   x 374 → 400   x 818 → 800
400,000   80,000   240,000

©1996 Kelley Wingate Publications   55   KW 1304

Worksheet 56

Name _____   Skill: Rounding and Estimating Numbers

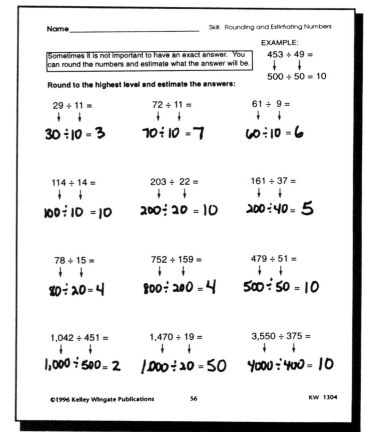

©1996 Kelley Wingate Publications   56   KW 1304

CD-3724

Answer Key

Skill: Logic

Use the ten clues below to find the correct digits in this "magic" number.

6 , 8 0 2 , 1 3 7 . 5 9 4

1. The millions digit is the difference of 15 and 9. **6**

2. The tenths digit is the quotient of 35 and 7. **5**

3. The tens digit is the difference of 11 and 8. **3**

4. The thousandths digit is the quotient of 20 and 5. **4**

5. The hundreds digit is the difference of 8 and 7. **1**

6. The hundred thousands digit is the product of 4 and 2. **8**

7. The thousands digit is the sum of 1 and 1. **2**

8. The ones digit is the quotient of 84 and 12. **7**

9. The hundredths digit is the difference of 16 and 7. **9**

10. The ten thousands digit is the product of 8 and 0. **0**

57 KW 1304

Skill: Logic

Use the ten clues below to find the correct digits in this "magic" number.

2 , 6 9 3 , 4 5 0 . 1 7 8

1. The hundreds digit is the difference of 13 and 9. **4**

2. The hundredths digit is the sum of 3 and 4. **7**

3. The tenths digit is the quotient of 5 and 5. **1**

4. The hundred thousands digit is the difference of 14 and 8. **6**

5. The thousandths digit is the sum of 5 and 3. **8**

6. The thousands digit is the quotient of 24 and 8. **3**

7. The tens digit is the difference of 13 and 8. **5**

8. The millions digit is the quotient of 22 and 11. **2**

9. The ones digit is the difference of 2 and 2. **0**

10. The ten thousands digit is the sum of 3 and 6. **9**

58 KW 1304

Skill: Logic

Use the ten clues below to find the correct digits in this "magic" number.

9 , 4 5 7 , 8 0 6 . 2 3 1

1. The tenths digit is the quotient of 18 and 9. **2**

2. The ten thousands digit is the difference of 12 and 7. **5**

3. The millions digit is the sum of 7 and 2. **9**

4. The hundredths digit is the quotient of 27 and 9. **3**

5. The ones digit is the difference of 13 and 7. **6**

6. The hundred thousands digit is the quotient of 32 and 8. **4**

7. The tens digit is the difference of 3 and 3. **0**

8. The thousands digit is the quotient of 84 and 12. **7**

9. The hundreds digit is the difference of 13 and 5. **8**

10. The thousandths digit is the quotient of 7 and 7. **1**

59 KW 1304

Skill: Logic

Use the ten clues below to find the correct digits in this "magic" number.

3 , 7 1 4 , 9 6 8 . 5 2 0

1. The thousandths digit is the product of 11 and 0. **0**

2. The hundred thousands digit is the quotient of 77 and 11. **7**

3. The hundreds digit is the difference of 11 and 2. **9**

4. The ones digit is the difference of 11 and 3. **8**

5. The hundredths digit is the quotient of 22 and 11. **2**

6. The millions digit is the product of 3 and 1. **3**

7. The thousands digit is the difference of 11 and 7. **4**

8. The tenths digit is the difference of 11 and 6. **5**

9. The tens digit is the quotient of 66 and 11. **6**

10. The ten thousands digit is the sum of 1 and 0. **1**

60 KW 1304

CD-3724

Answer Key

CD-3724

Name _____

Fraction - Ease

Fractions have **numerators** (top numbers) and **denominators** (bottom numbers).

When the denominators are the same you add or subtract the numerators. The denominator stays the same. Complete the problems below. The first one has been done for you.

$\frac{1}{3} + \frac{1}{3} = \boxed{\frac{2}{3}}$ $\frac{2}{5} - \frac{1}{5} = \boxed{\frac{1}{5}}$

$\frac{2}{7} + \frac{3}{7} = \boxed{\frac{5}{7}}$ $\frac{6}{11} - \frac{4}{11} = \boxed{\frac{2}{11}}$

$\frac{1}{8} + \frac{1}{8} = \boxed{\frac{2}{8}}$ $\frac{2}{9} + \frac{6}{9} = \boxed{\frac{8}{9}}$

$\frac{5}{9} - \frac{3}{9} = \boxed{\frac{2}{9}}$ $\frac{2}{5} + \frac{2}{5} = \boxed{\frac{4}{5}}$

$\frac{6}{7} - \frac{3}{7} = \boxed{\frac{3}{7}}$ $\frac{3}{4} - \frac{1}{4} = \boxed{\frac{2}{4}}$

$\frac{1}{2} - \frac{1}{2} = \boxed{\frac{0}{2}}\ 0$ $\frac{2}{6} + \frac{3}{6} = \boxed{\frac{5}{6}}$

$\frac{1}{5} + \frac{3}{5} = \boxed{\frac{4}{5}}$ $\frac{5}{7} - \frac{1}{7} = \boxed{\frac{4}{7}}$

$\frac{8}{9} - \frac{7}{9} = \boxed{\frac{1}{9}}$ $\frac{1}{10} + \frac{3}{10} = \boxed{\frac{4}{10}}$

$\frac{5}{6} - \frac{4}{6} = \boxed{\frac{1}{6}}$ $\frac{5}{8} + \frac{2}{8} = \boxed{\frac{7}{8}}$

61   KW 1304

Name _____

Fraction - Ease

Fractions have **numerators** (top numbers) and **denominators** (bottom numbers).

When the denominators are the same you add or subtract the numerators. The denominator stays the same. Complete the problems below. The first one has been done for you.

$\frac{2}{5} + \frac{1}{5} = \boxed{\frac{3}{5}}$ $\frac{8}{9} - \frac{3}{9} = \boxed{\frac{5}{9}}$

$\frac{7}{10} - \frac{3}{10} = \boxed{\frac{4}{10}}$ $\frac{3}{8} + \frac{4}{8} = \boxed{\frac{7}{8}}$

$\frac{5}{7} - \frac{4}{7} = \boxed{\frac{1}{7}}$ $\frac{3}{9} + \frac{4}{9} = \boxed{\frac{7}{9}}$

$\frac{5}{12} + \frac{6}{12} = \boxed{\frac{11}{12}}$ $\frac{2}{4} + \frac{1}{4} = \boxed{\frac{3}{4}}$

$\frac{9}{10} - \frac{1}{10} = \boxed{\frac{8}{10}}$ $\frac{3}{5} - \frac{2}{5} = \boxed{\frac{1}{5}}$

$\frac{4}{9} - \frac{2}{9} = \boxed{\frac{2}{9}}$ $\frac{1}{9} + \frac{2}{9} = \boxed{\frac{3}{9}}$

$\frac{1}{4} + \frac{3}{4} = \boxed{\frac{4}{4}}\ 1$ $\frac{7}{8} - \frac{4}{8} = \boxed{\frac{3}{8}}$

$\frac{4}{5} - \frac{1}{5} = \boxed{\frac{3}{5}}$ $\frac{4}{10} + \frac{5}{10} = \boxed{\frac{9}{10}}$

$\frac{6}{7} - \frac{4}{7} = \boxed{\frac{2}{7}}$ $\frac{1}{3} + \frac{1}{3} = \boxed{\frac{2}{3}}$

62   KW 1304

Name _____

Common Denominator

Write the first six multiples of each denominator in the pairs below. Underline or circle the first multiple that is the same for each pair. This number is the common denominator to be used when adding or subtracting the fractions.

A. $\frac{1}{3}$ 3, 6, 9, 12, 15   Common Denominator: ___6___
    $\frac{5}{6}$ 6, 12, 18, 24, 30

B. $\frac{2}{9}$ 9, 18, 27, 36, 45, 54   Common Denominator: 9
    $\frac{1}{3}$ 3, 6, 9, 12, 15, 18

C. $\frac{2}{3}$ 3, 6, 9, 12, 15, 18   Common Denominator: 12
    $\frac{1}{4}$ 4, 8, 12, 16, 20, 24

D. $\frac{2}{5}$ 5, 10, 15, 20, 25, 30   Common Denominator: 20
    $\frac{3}{4}$ 4, 8, 12, 16, 20, 24

E. $\frac{1}{2}$ 2, 4, 6, 8, 10, 12   Common Denominator: 6
    $\frac{2}{3}$ 3, 6, 9, 12, 15, 18

F. $\frac{7}{8}$ 8, 16, 24, 32, 40, 48   Common Denominator: 8
    $\frac{1}{2}$ 2, 4, 6, 8, 10, 12

63   KW 1304

Name _____

Common Denominator

Write the first seven multiples of each denominator in the pairs below. Underline or circle the first multiple that is the same for each pair. This number is the common denominator to be used when adding or subtracting fractions.

A. $\frac{2}{5}$ 5, 10, 15, 20, 25   Common Denominator: ___10___
    $\frac{7}{10}$ 10, 20, 30, 40, 50

B. $\frac{2}{3}$ 3, 6, 9, 12, 15, 18, 21   Common Denominator: 15
    $\frac{4}{5}$ 5, 10, 15, 20, 25, 30, 35

C. $\frac{7}{8}$ 8, 16, 24, 32, 40, 48, 56   Common Denominator: 24
    $\frac{1}{6}$ 6, 12, 18, 24, 30, 36, 42

D. $\frac{1}{2}$ 2, 4, 6, 8, 10, 12, 14   Common Denominator: 14
    $\frac{1}{7}$ 7, 14, 21, 28, 35, 42, 49

E. $\frac{1}{2}$ 2, 4, 6, 8, 10, 12, 14   Common Denominator: 4
    $\frac{1}{4}$ 4, 8, 12, 16, 20, 24, 28

F. $\frac{1}{3}$ 3, 6, 9, 12, 15, 18, 21   Common Denominator: 12
    $\frac{3}{4}$ 4, 8, 12, 16, 20, 24, 28

64   KW 1304

Answer Key

Name _____ Skill: Fractions

Compare Squares

Compare these fractions. Write >, <, or = in the square to make a true math statement. The first problem is done for you.

When fractions have the same denominator, compare the numerators.

$$\frac{1}{3} < \frac{2}{3}$$

(1 is less than 2)

When fractions have different denominators, find the common denominator.

$$\frac{1}{3} \quad \frac{1}{2}$$
$$\frac{2}{6} < \frac{3}{6}$$

4/12 $\frac{1}{3}$ < $\frac{2}{4}$ 6/12 $\frac{7}{8}$ > $\frac{3}{4}$ $\frac{8}{9}$ > $\frac{5}{6}$

$\frac{9}{10}$ > $\frac{8}{9}$ $\frac{3}{5}$ < $\frac{4}{5}$ $\frac{3}{7}$ < $\frac{4}{9}$

$\frac{7}{10}$ > $\frac{3}{5}$ $\frac{1}{4}$ < $\frac{3}{8}$ $\frac{4}{5}$ < $\frac{9}{10}$

$\frac{5}{6}$ < $\frac{7}{8}$ $\frac{1}{2}$ > $\frac{7}{11}$ $\frac{2}{3}$ < $\frac{3}{4}$

$\frac{2}{3}$ > $\frac{1}{3}$ $\frac{4}{5}$ < $\frac{7}{8}$ $\frac{5}{6}$ > $\frac{2}{3}$

$\frac{1}{6}$ < $\frac{2}{6}$ $\frac{4}{8}$ = $\frac{1}{2}$ $\frac{2}{5}$ < $\frac{3}{4}$

©1996 Kelley Wingate Publications   65   KW 1304

Name _____ Skill: Fractions

Compare Squares

Compare these fractions. Write >, <, or = in the square to make a true statement.

When fractions have different denominators, find the common denominator.

$$\frac{1}{3} \quad \frac{1}{2}$$
$$\frac{2}{6} < \frac{3}{6}$$

Any set of fractions may be compared by cross multiplying.

$$\frac{5}{8} \quad \frac{3}{6}$$
$$5 \times 6 = 30 > 8 \times 3 = 24$$

$\frac{5}{8}$ > $\frac{5}{9}$ $\frac{7}{14}$ = $\frac{1}{2}$ $\frac{3}{4}$ < $\frac{7}{9}$

$\frac{3}{5}$ > $\frac{7}{12}$ $\frac{3}{10}$ < $\frac{3}{4}$ $\frac{5}{6}$ > $\frac{3}{5}$

$\frac{1}{3}$ = $\frac{4}{12}$ $\frac{2}{6}$ = $\frac{4}{12}$ $\frac{2}{3}$ < $\frac{8}{9}$

$\frac{2}{5}$ > $\frac{1}{5}$ $\frac{6}{7}$ > $\frac{4}{5}$ $\frac{5}{7}$ < $\frac{3}{4}$

$\frac{4}{7}$ < $\frac{5}{8}$ $\frac{9}{11}$ < $\frac{9}{10}$ $\frac{2}{3}$ < $\frac{7}{8}$

$\frac{7}{8}$ > $\frac{10}{12}$ $\frac{6}{9}$ = $\frac{4}{6}$ $\frac{1}{4}$ < $\frac{5}{12}$

©1996 Kelley Wingate Publications   66   KW 1304

Name _____ Skill: Fractions

Fraction Features

Find a common denominator for each pair of fractions below then complete the problem.

1. $\frac{1}{2}$ − $\frac{1}{4}$

2. $\frac{5}{8}$ + $\frac{1}{4}$

3. $\frac{2}{3}$ + $\frac{1}{6}$

4. $\frac{5}{9}$ − $\frac{1}{3}$

5. $\frac{3}{4}$ − $\frac{1}{3}$

6. $\frac{9}{10}$ − $\frac{1}{2}$

7. $\frac{2}{6}$ − $\frac{1}{3}$

8. $\frac{5}{12}$ + $\frac{1}{2}$

9. $\frac{2}{5}$ + $\frac{4}{10}$

10. $\frac{11}{12}$ − $\frac{1}{6}$

11. $\frac{3}{14}$ + $\frac{1}{7}$

12. $\frac{1}{6}$ + $\frac{1}{2}$

13. $\frac{7}{8}$ − $\frac{1}{4}$

14. $\frac{2}{3}$ + $\frac{1}{5}$ $\frac{13}{15}$

15. $\frac{8}{9}$ − $\frac{2}{3}$ $\frac{13}{15}$

©1996 Kelley Wingate Publications   67   KW 1304

Name _____ Skill: Fractions

Fraction Features

Find a common denominator for each pair of fractions below then complete the problem.

1. $\frac{1}{6}$ = $\frac{2}{12}$ + $\frac{3}{12}$ = $\frac{3}{12}$ $\frac{11}{12}$

2. $\frac{7}{8}$ = $\frac{21}{24}$ − $\frac{1}{6}$ = $\frac{4}{24}$ $\frac{17}{24}$

3. $\frac{2}{3}$ = $\frac{10}{15}$ + $\frac{1}{5}$ = $\frac{3}{15}$ $\frac{13}{15}$

4. $\frac{7}{12}$ = $\frac{7}{12}$ − $\frac{1}{4}$ = $\frac{3}{12}$ $\frac{4}{12}$

5. $\frac{1}{4}$ = $\frac{5}{20}$ + $\frac{3}{5}$ = $\frac{12}{20}$ $\frac{17}{20}$

6. $\frac{3}{5}$ = $\frac{18}{30}$ − $\frac{1}{6}$ = $\frac{5}{30}$ $\frac{13}{30}$

7. $\frac{8}{9}$ = $\frac{16}{18}$ − $\frac{5}{6}$ = $\frac{15}{18}$ $\frac{1}{18}$

8. $\frac{1}{4}$ = $\frac{3}{12}$ + $\frac{1}{3}$ = $\frac{4}{12}$ $\frac{7}{12}$

9. $\frac{5}{12}$ = $\frac{25}{60}$ − $\frac{2}{10}$ = $\frac{12}{60}$ $\frac{13}{60}$

10. $\frac{1}{2}$ = $\frac{5}{10}$ + $\frac{1}{10}$ = $\frac{1}{10}$ $\frac{6}{10}$

11. $\frac{8}{9}$ = $\frac{8}{9}$ − $\frac{1}{3}$ = $\frac{3}{9}$ $\frac{5}{9}$

12. $\frac{1}{8}$ + $\frac{3}{4}$ $\frac{7}{8}$

13. $\frac{4}{7}$ = $\frac{20}{35}$ − $\frac{1}{5}$ = $\frac{7}{35}$ $\frac{13}{35}$

14. $\frac{9}{10}$ − $\frac{3}{5}$

15. $\frac{4}{9}$ = $\frac{16}{36}$ + $\frac{1}{4}$ = $\frac{9}{36}$ $\frac{25}{36}$

©1996 Kelley Wingate Publications   68   KW 1304

119

©1996 Kelley Wingate Publications   CD-3724

Answer Key

Worksheet (page 69)

Fraction Features

Find a common denominator for each pair of fractions below then complete the problem.

1. $\frac{4}{5} = \frac{16}{20}$
   $-\frac{3}{4} = \frac{15}{20}$
   $\frac{1}{20}$

2. $\frac{2}{3} = \frac{16}{24}$
   $+\frac{1}{8} = \frac{3}{24}$
   $\frac{19}{24}$

3. $\frac{5}{6} = \frac{15}{18}$
   $-\frac{2}{9} = \frac{4}{18}$
   $\frac{11}{18}$

4. $\frac{3}{8} = \frac{9}{24}$
   $+\frac{5}{12} = \frac{10}{24}$
   $\frac{19}{24}$

5. $\frac{1}{4} = \frac{3}{12}$
   $+\frac{1}{6} = \frac{2}{12}$
   $\frac{5}{12}$

6. $\frac{3}{4} = \frac{21}{28}$
   $-\frac{1}{7} = \frac{4}{28}$
   $\frac{17}{28}$

7. $\frac{1}{2} = \frac{7}{14}$
   $-\frac{1}{14} = \frac{1}{14}$
   $\frac{6}{14}$

8. $\frac{7}{12} = \frac{7}{12}$
   $+\frac{1}{3} = \frac{4}{12}$
   $\frac{11}{12}$

9. $\frac{7}{8} = \frac{21}{24}$
   $-\frac{5}{12} = \frac{10}{24}$
   $\frac{11}{24}$

10. $\frac{1}{4} = \frac{9}{36}$
    $+\frac{2}{9} = \frac{8}{36}$
    $\frac{17}{36}$

11. $\frac{3}{7} = \frac{9}{21}$
    $-\frac{1}{3} = \frac{7}{21}$
    $\frac{2}{21}$

12. $\frac{3}{10} = \frac{3}{10}$
    $+\frac{1}{10} = \frac{1}{10}$
    $\frac{5}{10}$

13. $\frac{5}{11} = \frac{10}{22}$
    $+\frac{1}{2} = \frac{11}{22}$
    $\frac{21}{22}$

14. $\frac{2}{3} = \frac{14}{21}$
    $-\frac{2}{7} = \frac{6}{21}$
    $\frac{8}{21}$

15. $\frac{5}{8} = \frac{5}{8}$
    $-\frac{2}{4} = \frac{4}{8}$
    $\frac{1}{8}$

Worksheet (page 70)

Fraction Features

Find a common denominator for each pair of fractions below then complete the problem.

1. $\frac{3}{4} = \frac{9}{12}$
   $-\frac{2}{3} = \frac{8}{12}$
   $\frac{1}{12}$

2. $\frac{5}{6} = \frac{15}{18}$
   $+\frac{1}{9} = \frac{2}{18}$
   $\frac{17}{18}$

3. $\frac{1}{8} = \frac{3}{24}$
   $+\frac{2}{3} = \frac{16}{24}$
   $\frac{19}{24}$

4. $\frac{3}{4} = \frac{9}{12}$
   $-\frac{1}{6} = \frac{2}{12}$
   $\frac{7}{12}$

5. $\frac{3}{7} = \frac{6}{14}$
   $+\frac{1}{2} = \frac{7}{14}$
   $\frac{13}{14}$

6. $\frac{2}{5} = \frac{6}{15}$
   $-\frac{1}{3} = \frac{5}{15}$
   $\frac{1}{15}$

7. $\frac{2}{3} = \frac{8}{12}$
   $+\frac{1}{4} = \frac{3}{12}$
   $\frac{11}{12}$

8. $\frac{3}{4} = \frac{3}{4}$
   $-\frac{1}{2} = \frac{2}{4}$
   $\frac{1}{4}$

9. $\frac{1}{9} = \frac{1}{9}$
   $+\frac{1}{3} = \frac{3}{9}$
   $\frac{4}{9}$

10. $\frac{15}{16} = \frac{15}{16}$
    $-\frac{1}{4} = \frac{4}{16}$
    $\frac{11}{16}$

11. $\frac{1}{15} = \frac{1}{15}$
    $+\frac{1}{5} = \frac{3}{15}$
    $\frac{4}{15}$

12. $\frac{5}{6} = \frac{15}{18}$
    $-\frac{7}{9} = \frac{14}{18}$
    $\frac{1}{18}$

13. $\frac{6}{7} = \frac{18}{21}$
    $-\frac{1}{3} = \frac{7}{21}$
    $\frac{11}{21}$

14. $\frac{2}{8} = \frac{10}{40}$
    $+\frac{3}{5} = \frac{24}{40}$
    $\frac{34}{40}$

15. $\frac{7}{8} = \frac{7}{8}$
    $-\frac{1}{4} = \frac{2}{8}$
    $\frac{5}{8}$

Worksheet (page 71)

Decimal Dimensions

To add or subtract decimals, line up the decimal points and fill in any blank spaces with a zero. Add or subtract as usual. Do not forget to bring the decimal point straight down into your answer.

EXAMPLES:

| 1.6
+ 33.45 | 01.60
+ 33.45 | 01.60
+ 33.45
35.05 | 24.8
- 13.75 | 24.80
- 13.75 | 24.80
- 13.75
21.05 |

1) $6.01 + 83.467 = 89.477$

2) $7 - 3.21 = 3.79$

3) $13.3 - 8.64 = 4.66$

4) $1.9 + 3.87 + 5.0 = 10.77$

5) $10.35 - 1.728 = 8.622$

6) $43.8 + 71.06 = 114.86$

7) $18.5 - 2.391 = 16.109$

8) $9.2 - 4.89 = 4.31$

9) $3.36 + 14.2 = 17.56$

10) $157.3 + 88.29 = 245.59$

11) $15.0 - 6.781 = 8.219$

12) $153.5 + 19.8 + 7.32 = 180.62$

Worksheet (page 72)

Decimal Dimensions

To add or subtract decimals, line up the decimal points and fill in any blank spaces with a zero. Add or subtract as usual. Do not forget to bring the decimal point straight down into your answer.

EXAMPLES:

| 1.6
+ 33.45 | 01.60
+ 33.45 | 01.60
+ 33.45
35.05 | 24.8
- 13.75 | 24.80
- 13.75 | 24.80
- 13.75
11.05 |

1) $6.25 + 14.0 + 36.75 = 57.00$

2) $4.3 - 2.007 = 2.293$

3) $91.8 - 12.17 = 79.63$

4) $5.5 + 3.03 + 0.004 = 8.543$

5) $116.09 - 97.387 = 18.703$

6) $19.82 - 7.002 = 12.818$

7) $49.7 - 39.95 = 9.75$

8) $5.2 - 3.84 = 1.36$

9) $0.51 + 3.60 + 14.00 = 18.11$

10) $17.0 + 3.9 + 1.75 = 22.65$

11) $440.01 - 289.607 = 150.403$

12) $40.0 - 38.81 = 1.19$

Answer Key

Decimal Dimensions

To multiply decimals: 1) Multiply as usual; 2) In the final product, move the decimal once to the left for every decimal place in the original two multipliers (numbers multiplied together)

EXAMPLES:

1.5	1.5	1 place		8.3	8.3	1 place
x 0.3	x 0.3	1 place		x .002	x .002	3 places
045	0.45	2 places		00166	0.0166	4 places

1) 0.4 x 0.8 **.32**	2) 1.6 x 3 **4.8**	3) 21.2 x 3.3 **69.96**
4) 7.01 x 0.4 **2.804**	5) 9.8 x 0.3 **2.94**	6) .004 x .5 **.002**
7) 1.23 x 5.4 **6.642**	8) 28.2 x .4 **11.28**	9) 4.97 x 6 **29.82**
10) 5.8 x .3 **1.74**	11) 2.5 x 3.7 **9.25**	12) 5.08 x 7.2 **36.576**

©1996 Kelley Wingate Publications   73   KW 1304

Decimal Dimensions

To multiply decimals: 1)Multiply as usual; 2) In the final product, move the decimal once to the left for every decimal place in the original two multipliers (numbers multiplied together)

EXAMPLES:

1.5	1.5	1 place		8.3	8.3	1 place
x 0.3	x 0.3	1 place		x .002	x .002	3 places
045	0.45	2 places		00166	0.0166	4 places

1) 1.8 x 4 **7.2**	2) 7.3 x 8 **58.4**	3) 5.05 x 3 **15.15**
4) 4.43 x 7 **31.01**	5) 4.8 x 2.3 **11.04**	6) 5.2 x 8.9 **46.28**
7) 27.4 x 1.7 **46.58**	8) 38.4 x 3.2 **122.88**	9) .84 x 6.7 **5.628**
10) 3.01 x 4.9 **14.749**	11) 5.003 x 1.4 **7.0042**	12) .006 x .2 **.0012**

©1996 Kelley Wingate Publications   74   KW 1304

Measure Sense
Standard

Length
1 inch (in)
12 ins = 1 foot (ft)
3 ft = 1 yard (yd)
5,280 ft = 1 mile (mi)

1. length of a football field **(100 yd.)** 100 mi.	2. height of a doorway 7 in. **(7 ft.)**	3. width of a math book **(8 in.)** 8 ft.
4. height of a desk 2 in. **(2 ft.)**	5. length of a swimming pool **(36 ft.)** 36 mi.	6. width of your wrist **(3 in.)** 3 ft.
7. length of your foot **(7 in.)** 7 ft.	8. length of a car **(100 in.)** 100 ft.	9. height of a fence 4 in. **(4 ft.)**
10. your house to the corner **(50 yd.)** 50 mi.	11. height of a basketball player **(7 ft.)** 7 yd.	12. height of a house **(20 ft.)** 20 yd.

©1996 Kelley Wingate Publications   75   KW 1304

Measure Sense
Metric

Length	
Millimeter (mm): thickness of a coin	10 mm. = 1 cm.
Centimeter (cm): one fingernail	10 cm. = 1 dm.
Decimeter (dm): height of a small can	10 dm. = 1 m.
Meter (m): lenth of a baseball bat	1000 m. = 1 km.
Kilometer (km): about 2/3 of a mile	

1. height of a wall in your home 3 mm. **(3 m.)**	2. piece of notebook paper **(15 cm.)** 15 dm.	3. thickness of an envelope **(1 mm.)** 1 cm.
4. length of a football 3 cm. **(3 dm.)**	5. width of a dictionary 7 mm. **(7 cm.)**	6. height of a fence 2 cm. **(2 m.)**
7. liter of soda **(30 cm.)** 30 m.	8. height of a chair 1 dm. **(1 m.)**	9. from one city to another 40 m. **(40 km.)**
10. your house to your school 2 m. **(2 km.)**	11. width of your desk 4 mm. **(4 dm.)**	12. thickness of a computer disk **(1 mm.)** 1 dm.

©1996 Kelley Wingate Publications   76   KW 1304

Answer Key

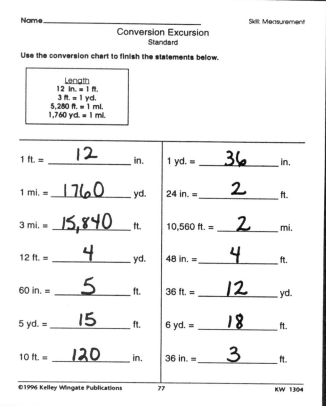

Name_____ Skill: Measurement

Conversion Excursion
Standard

Use the conversion chart to finish the statements below.

Length
12 in. = 1 ft.
3 ft. = 1 yd.
5,280 ft. = 1 mi.
1,760 yd. = 1 mi.

1 ft. = **12** in. | 1 yd. = **36** in.

1 mi. = **1760** yd. | 24 in. = **2** ft.

3 mi. = **15,840** ft. | 10,560 ft. = **2** mi.

12 ft. = **4** yd. | 48 in. = **4** ft.

60 in. = **5** ft. | 36 ft. = **12** yd.

5 yd. = **15** ft. | 6 yd. = **18** ft.

10 ft. = **120** in. | 36 in. = **3** ft.

©1996 Kelley Wingate Publications   77   KW 1304

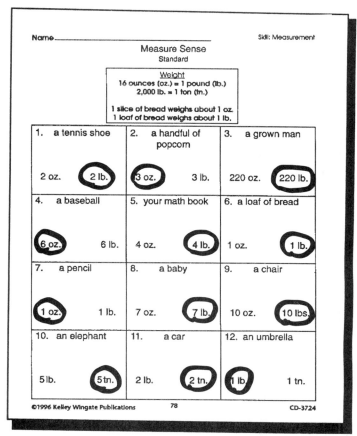

Name_____ Skill: Measurement

Measure Sense
Standard

Weight
16 ounces (oz.) = 1 pound (lb.)
2,000 lb. = 1 ton (tn.)
1 slice of bread weighs about 1 oz.
1 loaf of bread weighs about 1 lb.

1. a tennis shoe	2. a handful of popcorn	3. a grown man
2 oz. (2 lb.)	(3 oz.) 3 lb.	220 oz. (220 lb.)
4. a baseball	5. your math book	6. a loaf of bread
(6 oz.) 6 lb.	4 oz. (4 lb.)	1 oz. (1 lb.)
7. a pencil	8. a baby	9. a chair
(1 oz.) 1 lb.	7 oz. (7 lb.)	10 oz. (10 lbs.)
10. an elephant	11. a car	12. an umbrella
5 lb. (5 tn.)	2 lb. (2 tn.)	(1 lb.) 1 tn.

©1996 Kelley Wingate Publications   78   CD-3724

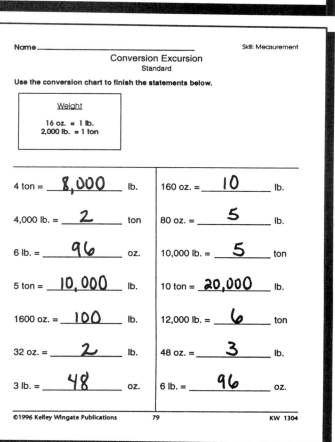

Name_____ Skill: Measurement

Conversion Excursion
Standard

Use the conversion chart to finish the statements below.

Weight
16 oz. = 1 lb.
2,000 lb. = 1 ton

4 ton = **8,000** lb. | 160 oz. = **10** lb.

4,000 lb. = **2** ton | 80 oz. = **5** lb.

6 lb. = **96** oz. | 10,000 lb. = **5** ton

5 ton = **10,000** lb. | 10 ton = **20,000** lb.

1600 oz. = **100** lb. | 12,000 lb. = **6** ton

32 oz. = **2** lb. | 48 oz. = **3** lb.

3 lb. = **48** oz. | 6 lb. = **96** oz.

©1996 Kelley Wingate Publications   79   KW 1304

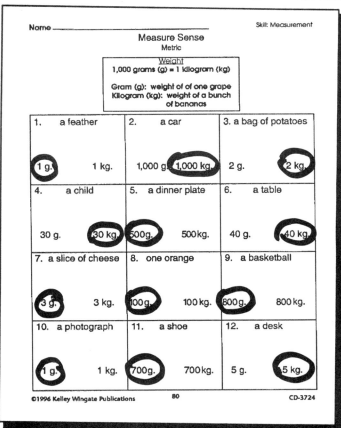

Name_____ Skill: Measurement

Measure Sense
Metric

Weight
1,000 grams (g) = 1 kilogram (kg)
Gram (g): weight of of one grape
Kilogram (kg): weight of a bunch of bananas

1. a feather	2. a car	3. a bag of potatoes
(1 g.) 1 kg.	1,000 g. (1,000 kg.)	2 g. (2 kg.)
4. a child	5. a dinner plate	6. a table
30 g. (30 kg.)	(500 g.) 500 kg.	40 g. (40 kg.)
7. a slice of cheese	8. one orange	9. a basketball
(3 g.) 3 kg.	(100 g.) 100 kg.	(800 g.) 800 kg.
10. a photograph	11. a shoe	12. a desk
(1 g.) 1 kg.	(700 g.) 700 kg.	5 g. (5 kg.)

©1996 Kelley Wingate Publications   80   CD-3724

122

CD-3724

Answer Key

Measure Sense — Standard

Skill: Measurement

Capacity
1 cup (c)
2 c = 1 pint (pt)
2 pts = 1 quart (qt)
4 qts = 1 gallon (gal)

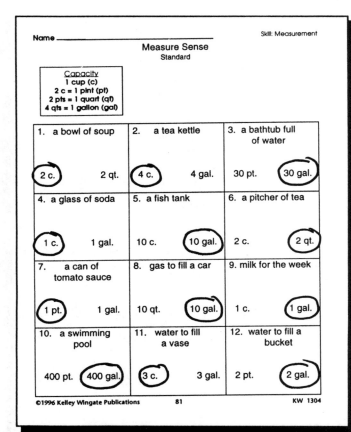

1. a bowl of soup	2. a tea kettle	3. a bathtub full of water
(2 c.) 2 qt.	(4 c.) 4 gal.	30 pt. (30 gal.)
4. a glass of soda	5. a fish tank	6. a pitcher of tea
(1 c.) 1 gal.	10 c. (10 gal.)	2 c. (2 qt.)
7. a can of tomato sauce	8. gas to fill a car	9. milk for the week
(1 pt.) 1 gal.	10 qt. (10 gal.)	1 c. (1 gal.)
10. a swimming pool	11. water to fill a vase	12. water to fill a bucket
400 pt. (400 gal.)	(3 c.) 3 gal.	2 pt. (2 gal.)

©1996 Kelley Wingate Publications   81   KW 1304

Measure Sense — Metric

Skill: Measurement

Capacity
1,000 mL. = 1 liter (L)

Milliliter (ml) = eye dropper full
Liter (l) = half of a 2 liter bottle of soda

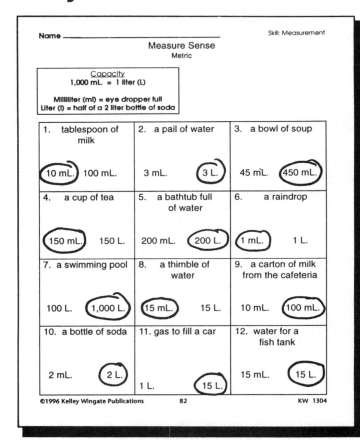

1. tablespoon of milk	2. a pail of water	3. a bowl of soup
(10 mL.) 100 mL.	3 mL. (3 L.)	45 mL. (450 mL.)
4. a cup of tea	5. a bathtub full of water	6. a raindrop
(150 mL.) 150 L.	200 mL. (200 L.)	(1 mL.) 1 L.
7. a swimming pool	8. a thimble of water	9. a carton of milk from the cafeteria
100 L. (1,000 L.)	(15 mL.) 15 L.	10 mL. (100 mL.)
10. a bottle of soda	11. gas to fill a car	12. water for a fish tank
2 mL. (2 L.)	1 L. (15 L.)	15 mL. (15 L.)

©1996 Kelley Wingate Publications   82   KW 1304

Conversion Excursion — Standard

Skill: Measurement

Use the conversion chart to finish the statements below.

Capacity
2 c. = 1 pt
2 pt. = 1 qt.
4 qt. = 1 gal

8 pt. = **4** qt.   6 gal. = **24** qt.

20 c. = **10** pt.   4 pt. = **8** c.

4 qt. = **8** pt.   14 c. = **7** pt.

32 qt. = **8** gal.   10 c. = **5** pt.

12 pt. = **6** qt.   4 gal. = **32** pt.

16 qt. = **4** gal.   5 qt. = **20** c.

32 c. = **8** qt.   3 qt. = **6** pt.

©1996 Kelley Wingate Publications   83   CD-3724

Measure Sense — Celsius (C) and Fahrenheit (F)

Temperature		
Fahrenheit (F) Celsius (C)		
Water freezes	32° F	0° C
Water boils	212° F	100° C

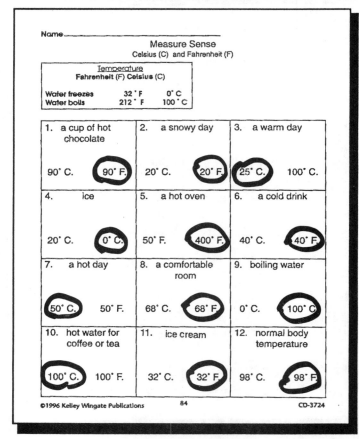

1. a cup of hot chocolate	2. a snowy day	3. a warm day
90° C. (90° F.)	20° C. (20° F.)	(25° C.) 100° C.
4. ice	5. a hot oven	6. a cold drink
20° C. (0° C.)	50° F. (400° F.)	40° C. (40° F.)
7. a hot day	8. a comfortable room	9. boiling water
(50° C.) 50° F.	68° C. (68° F.)	0° C. (100° C.)
10. hot water for coffee or tea	11. ice cream	12. normal body temperature
(100° C.) 100° F.	32° C. (32° F.)	98° C. (98° F.)

©1996 Kelley Wingate Publications   84   CD-3724

©1996 Kelley Wingate Publications   CD-3724

Answer Key

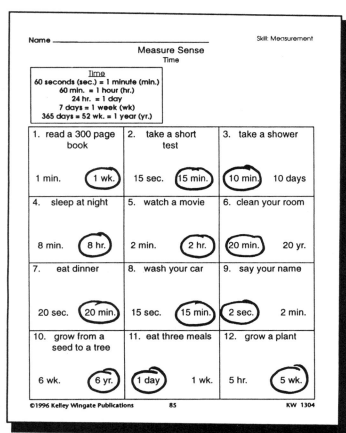

Name _____   Skill: Measurement
Measure Sense
Time

Time
60 seconds (sec.) = 1 minute (min.)
60 min. = 1 hour (hr.)
24 hr. = 1 day
7 days = 1 week (wk)
365 days = 52 wk. = 1 year (yr.)

1. read a 300 page book	2. take a short test	3. take a shower
1 min. (1 wk.)	15 sec. (15 min.)	(10 min.) 10 days
4. sleep at night	**5.** watch a movie	**6.** clean your room
8 min. (8 hr.)	2 min. (2 hr.)	(20 min.) 20 yr.
7. eat dinner	**8.** wash your car	**9.** say your name
20 sec. (20 min.)	15 sec. (15 min.)	(2 sec.) 2 min.
10. grow from a seed to a tree	**11.** eat three meals	**12.** grow a plant
6 wk. (6 yr.)	(1 day) 1 wk.	5 hr. (5 wk.)

©1996 Kelley Wingate Publications   85   KW 1304

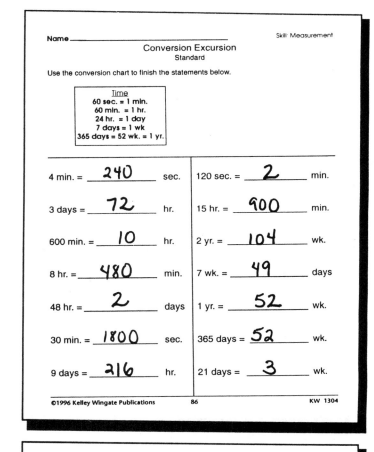

Name _____   Skill: Measurement
Conversion Excursion
Standard

Use the conversion chart to finish the statements below.

Time
60 sec. = 1 min.
60 min. = 1 hr.
24 hr. = 1 day
7 days = 1 wk
365 days = 52 wk. = 1 yr.

4 min. = **240** sec.	120 sec. = **2** min.
3 days = **72** hr.	15 hr. = **900** min.
600 min. = **10** hr.	2 yr. = **104** wk.
8 hr. = **480** min.	7 wk. = **49** days
48 hr. = **2** days	1 yr. = **52** wk.
30 min. = **1800** sec.	365 days = **52** wk.
9 days = **216** hr.	21 days = **3** wk.

©1996 Kelley Wingate Publications   86   KW 1304

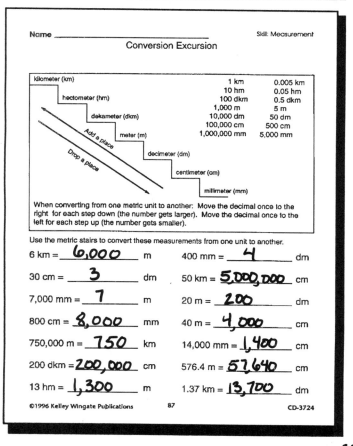

Name _____   Skill: Measurement
Conversion Excursion

	1 km	0.005 km
kilometer (km)	10 hm	0.05 hm
hectometer (hm)	100 dkm	0.5 dkm
dekameter (dkm)	1,000 m	5 m
meter (m)	10,000 dm	50 dm
decimeter (dm)	100,000 cm	500 cm
centimeter (cm)	1,000,000 mm	5,000 mm
millimeter (mm)		

Add a place / Drop a place

When converting from one metric unit to another: Move the decimal once to the right for each step down (the number gets larger). Move the decimal once to the left for each step up (the number gets smaller).

Use the metric stairs to convert these measurements from one unit to another.

6 km = **6,000** m	400 mm = **4** dm
30 cm = **3** dm	50 km = **5,000,000** cm
7,000 mm = **7** m	20 m = **200** dm
800 cm = **8,000** mm	40 m = **4,000** cm
750,000 m = **750** km	14,000 mm = **1,400** cm
200 dkm = **200,000** cm	576.4 m = **57,640** cm
13 hm = **1,300** m	1.37 km = **13,700** dm

©1996 Kelley Wingate Publications   87   CD-3724

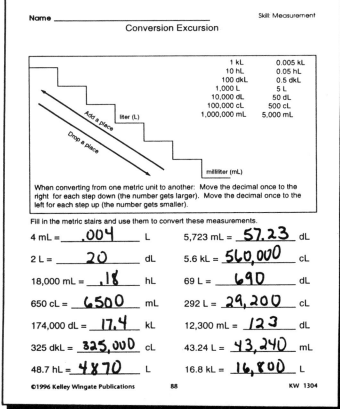

Name _____   Skill: Measurement
Conversion Excursion

	1 kL	0.005 kL
	10 hL	0.05 hL
	100 dkL	0.5 dkL
	1,000 L	5 L
liter (L)	10,000 dL	50 dL
	100,000 cL	500 cL
	1,000,000 mL	5,000 mL
milliliter (mL)		

Add a place / Drop a place

When converting from one metric unit to another: Move the decimal once to the right for each step down (the number gets larger). Move the decimal once to the left for each step up (the number gets smaller).

Fill in the metric stairs and use them to convert these measurements.

4 mL = **.004** L	5,723 mL = **57.23** dL
2 L = **20** dL	5.6 kL = **560,000** cL
18,000 mL = **.18** hL	69 L = **690** dL
650 cL = **6500** mL	292 L = **29,200** cL
174,000 dL = **17.4** kL	12,300 mL = **123** dL
325 dkL = **325,000** cL	43.24 L = **43,240** mL
48.7 hL = **4870** L	16.8 kL = **16,800** L

©1996 Kelley Wingate Publications   88   KW 1304

Answer Key

Worksheet 1 (page 89)

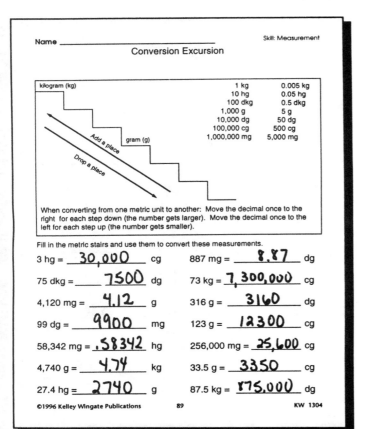

Skill: Measurement

Conversion Excursion

kilogram (kg)

1 kg	0.005 kg
10 hg	0.05 hg
100 dkg	0.5 dkg
1,000 g	5 g
10,000 dg	50 dg
100,000 cg	500 cg
1,000,000 mg	5,000 mg

Add a place
gram (g)
Drop a place

When converting from one metric unit to another: Move the decimal once to the right for each step down (the number gets larger). Move the decimal once to the left for each step up (the number gets smaller).

Fill in the metric stairs and use them to convert these measurements.

3 hg = **30,000** cg   887 mg = **8.87** dg

75 dkg = **7500** dg   73 kg = **7,300,000** cg

4,120 mg = **4.12** g   316 g = **3160** dg

99 dg = **9900** mg   123 g = **12300** cg

58,342 mg = **.58342** hg   256,000 mg = **25,600** cg

4,740 g = **4.74** kg   33.5 g = **3350** cg

27.4 hg = **2740** g   87.5 kg = **875,000** dg

©1996 Kelley Wingate Publications   89   KW 1304

Worksheet 2 (page 90)

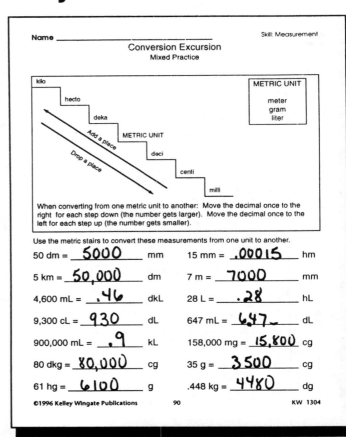

Name _____

Skill: Measurement

Conversion Excursion
Mixed Practice

kilo
hecto
deka
METRIC UNIT
Add a place
deci
Drop a place
centi
milli

| METRIC UNIT |
| meter |
| gram |
| liter |

When converting from one metric unit to another: Move the decimal once to the right for each step down (the number gets larger). Move the decimal once to the left for each step up (the number gets smaller).

Use the metric stairs to convert these measurements from one unit to another.

50 dm = **5000** mm   15 mm = **.00015** hm

5 km = **50,000** dm   7 m = **7000** mm

4,600 mL = **.46** dkL   28 L = **.28** hL

9,300 cL = **930** dL   647 mL = **6.47** dL

900,000 mL = **.9** kL   158,000 mg = **15,800** cg

80 dkg = **80,000** cg   35 g = **3500** cg

61 hg = **6100** g   .448 kg = **4480** dg

©1996 Kelley Wingate Publications   90   KW 1304

Worksheet 3 (page 91)

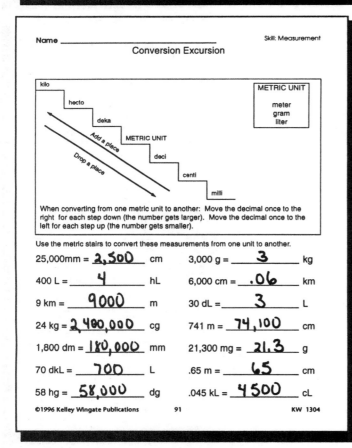

Name _____

Skill: Measurement

Conversion Excursion

kilo
hecto
deka
METRIC UNIT
Add a place
deci
Drop a place
centi
milli

| METRIC UNIT |
| meter |
| gram |
| liter |

When converting from one metric unit to another: Move the decimal once to the right for each step down (the number gets larger). Move the decimal once to the left for each step up (the number gets smaller).

Use the metric stairs to convert these measurements from one unit to another.

25,000 mm = **2,500** cm   3,000 g = **3** kg

400 L = **4** hL   6,000 cm = **.06** km

9 km = **9000** m   30 dL = **3** L

24 kg = **2,400,000** cg   741 m = **74,100** cm

1,800 dm = **180,000** mm   21,300 mg = **21.3** g

70 dkL = **700** L   .65 m = **65** cm

58 hg = **58,000** dg   .045 kL = **4500** cL

©1996 Kelley Wingate Publications   91   KW 1304

Worksheet 4 (page 92)

Name _____

Skill: Word Problems

Words Into Math

FOOD	VOTES
Tacos	★★★★★
Chicken Nuggets	★★★★
Pizza	★★★★★★★★★★
Hot Dog	★★★★★
Hamburger	★★★★★★★
Spaghetti	★★★

Jessica and Erica did a survey for a math project. They charted the favorite cafeteria foods of their whole 4th grade math class. Each star stands for one vote.

1. Which food do the students like best?
Pizza

2. How many students are in this 4th grade math class?
38

3. How many more children voted for tacos than spaghetti?
3

4. How many students voted for pizza or hotdogs?
17

5. Which food did Erica vote for?
don't know

6. How many more students voted for pizza than spaghetti?
8

7. What class gets to vote next week?
don't know

8. Which food got the fewest votes?
spaghetti

9. Which two foods got the same number of votes?
Tacos/hotdogs

10. How many students wanted to add red beans and rice to the menu?
don't know

©1996 Kelley Wingate Publications   92   KW 1304

125

Answer Key

Name _____

Words Into Math

Skill: Word Problems

Daily Schedule	TIME
Homeroom	8:00 - 8:15
1st Period	8:15 - 9:15
2nd Period	9:20 - 10:20
3rd Period	10:25 - 11:25
Lunch	11:25 - 12:00
4th Period	12:00 - 1:00
5th Period	1:05 - 2:05
6th Period	2:10 - 3:10

Charles High School has this daily schedule. It names all of the class periods and gives the time each one starts and ends. Use this schedule to answer the questions below.

1. How long is each class period?
1 hour

2. If Jimmy gets to class at 8:30, how much of 1st period did he miss?
15 mins.

3. What class does Susan have 4th Period?
don't know

4. Kaitlin's mom picked her up at 2:00. During what period did she leave?
5th

5. Jamal stays in Mr. Woods room for 2nd and 3rd period. How long is he in that room?
2 hours 5 mins.

6. What period does Ms. Smith teach gym?
don't know

7. How long is the lunch period?
35 mins.

8. Which period begins at 9:20?
2nd

9. What event happens between 8 and 8:15 each morning?
Homeroom

10. If school is over after 6th period, what time do the students get out?
3:10

Name _____

Words Into Math

Skill: Word Problems

Read the paragraph carefully then answer the questions.

Mrs. Gros's 4th grade class is going on a field trip to the zoo. There are 25 students in her class. Each child must pay $2.00 to get in the zoo and $1.25 for the bus fare. The class left the school at 9:00 and returned at 2:20.

1. What was the total each student paid to go on this trip?
$3.25

2. How much money did Mrs. Gros collect from the class all together?
$81.25

3. One parent for every 5 students came along on the trip. How many parents came?
5

4. Did Daniel like the sharks or the polar bears better?
don't know

5. How long was the class gone from school?
5 hrs. 20 mins.

6. How much did the zoo collect for all 25 students?
$50

7. How much more did it cost to get in the zoo than to ride the bus?
$.75

8. Did the students behave nicely on this trip?
don't know

Name _____

Words Into Math

Skill: Word Problems

At Mr. Lopez's sporting goods store, these baseball items are on sale this week.

1. Javier bought a new bat and three balls. How much did he spend?
$28.96

2. Eric has $2.00. How many packs of trading cards can he buy?
3

3. What is the cost of a mitt and a hat together?
$21.98

4. Kara has $40.00. Can she buy the shoes and a mitt or just one of them?
both

5. How many baseballs can Carlos get for his $10.00?
3

6. Does the store have Jenny's size shoes?
don't know

7. Which costs more: a bat and mitt, or shoes and a hat?
a bat and cards

8. How many more hits did Matthew get after he used his new bat?
don't know

9. Which costs more: one hat, two baseballs, or ten packs of trading cards?
a hat

10. Round each item to the nearest dollar. How much would it cost to get one of everything?
$34.00

Name _____

Words Into Math

Skill: Word Problems

Mrs. Jacob put up a chart to show how many canned goods her students collected for the food drive. Use the chart to answer the questions.

Students Canned goods					
Nicole	♥	Halie	♥♥♥	Kara	♥♥♥♥
Miriam	♥♥	Josh	♥♥	Eric	♥♥♥♥
Javier	♥♥♥♥	Catherine	♥	Kaitlin	♥♥♥
Charlie	♥	Jimmy	♥♥	Matt	♥♥♥♥♥♥
Tina	♥	Ashley	♥♥	Lindsey	♥
Tommy	♥	Brandon	♥♥	Jesse	♥♥

♥ = 1 can   ♥ = 2 cans

1. Which student brought in the most cans?
matt

2. Which students brought in the fewest cans?
Charlie + Tina

3. How many students brought in three or less cans?
7

4. How many students are in Mrs. Jacob's class?
18

5. Is Jesse a boy or a girl?
don't know

6. How many cans did the class collect all together?
76 cans

7. How many more cans did Matt bring than Josh?
7

8. How many cans did Kaitlin, Kara, and Catherine bring in together?
15

9. What kind of food was in Charlie's can?
don't know

10. Who brought in more: Eric and Ashley, or Javier and Miriam?
Javier + Miriam

Answer Key

127

Worksheet 1 (page 97)

Name _____ Skill: Review

Skills Evaluation

Choose the best answer to these review questions. Circle the correct answer.

1. Add: 43,442 + 24,758
- A. 67,100
- B. 67,390
- C. 68,190
- **D. 68,200** (circled)

2. Subtract: 3,047 − 1,869
- A. 1,078
- **B. 1,178** (circled)
- C. 2,822
- D. 2,812

3. Subtract: 85,040 − 43,639
- A. 41,419
- B. 42,679
- **C. 41,401** (circled)
- D. 42,411

4. Multiply: 392 x 47
- **A. 18,424** (circled)
- B. 14,862
- C. 27,394
- D. 18,433

5. Which of these expressions does not equal 9?
- A. 17 − 8
- B. 63 ÷ 7
- **C. 40 ÷ 5** (circled)
- D. 14 − 5

6. Which of these expressions does not equal 6?
- **A. 24 ÷ 6** (circled)
- B. 15 − 9
- C. 72 ÷ 12
- D. 3 x 2

7. Solve: 3 x 8 ÷ 6 = ☐
- A. 6
- B. 5
- **C. 4** (circled)
- D. 3

8. Solve: 6 x 6 ÷ 9 = ☐
- A. 3
- **B. 4** (circled)
- C. 5
- D. 6

9. Divide: 324 ÷ 6 = ☐
- A. 204
- B. 64
- C. 53
- **D. 54** (circled)

10. Divide: 504 ÷ 8 = ☐
- **A. 63** (circled)
- B. 86
- C. 74
- D. 52

©1996 Kelley Wingate Publications   97   KW 1304

Worksheet 2 (page 98)

Name _____ Skill: Review

Skills Evaluation

Choose the best answer to these review questions. Circle the correct answer.

1. Add: $\frac{1}{4} + \frac{1}{3}$
- A. 2/7
- B. 2/4
- **C. 7/12** (circled)
- D. 4/3

2. Add: $\frac{2}{5} + \frac{3}{6}$
- A. 5/11
- B. 5/14
- C. 5/30
- **D. 27/30** (circled)

3. The number is 3,157,246. The 3 is in which place?
- A. hundreds
- B. thousands
- C. thousandths
- **D. millions** (circled)

4. The number is 403,521.78. The 8 is in which place?
- A. ones
- B. tenths
- C. thousandths
- **D. hundredths** (circled)

5. What is the least common denominator for 5/6 and 2/9?
- A. 54
- **B. 18** (circled)
- C. 36
- D. 9

6. What is the least common denominator for 4/5 and 7/8?
- A. 16
- B. 32
- **C. 40** (circled)
- D. 45

7. Round to the nearest tenth: 33.765
- A. 30
- B. 34
- C. 33.7
- **D. 33.8** (circled)

8. Round to the nearest hundred: 32,172
- A. 32,170
- **B. 32,200** (circled)
- C. 32,000
- D. 30,000

9. Round to the nearest ten thousand: 154,998
- A. 154,000
- B. 155,000
- **C. 150,000** (circled)
- D. 160,000

10. Round to the nearest million: 75,523,479
- A. 75,000,000
- **B. 76,000,000** (circled)
- C. 70,000,000
- D. 80,000,000

©1996 Kelley Wingate Publications   98   KW 1304

Worksheet 3 (page 99)

Name _____ Skill: Review

Skills Evaluation

Choose the best answer to these review questions. Circle the correct answer.

1. Tell if the expression is <, >, or =. 72 ÷ 8 ☐ 15 − 9
- **A. >** (circled)
- B. <
- C. =

2. Tell if the expression is <, >, or =. 12 x 3 ☐ 4 x 9
- A. >
- B. <
- **C. =** (circled)

3. Tell if the expression is <, >, or =. 344 ÷ 8 ☐ 5 x 9
- **A. >** (circled)
- B. <
- C. =

4. Tell if the expression is <, >, or =. 27 + 31 ☐ 6 x 9
- **A. >** (circled)
- B. <
- C. =

5. Add: 1.06 + 3.9 + 5
- A. 8.69
- **B. 9.96** (circled)
- C. 1.50
- D. 4.01

6. Subtract: 5 − 3.789
- A. 3.784
- **B. 1.211** (circled)
- C. 3.794
- D. 1.221

7. Multiply: 452 x 7
- A. 2,852
- B. 2,862
- C. 3,064
- **D. 3,164** (circled)

8. Multiply: 1640 x 62
- A. 10,168
- B. 3,280
- **C. 101,680** (circled)
- D. 98,480

9. Multiply: 4.22 x 3.6
- **A. 15.192** (circled)
- B. 1.5192
- C. 151.92
- D. 1519.2

10. Multiply: 35.8 x .034
- A. 121.7200
- B. 12.1720
- **C. 1.2172** (circled)
- D. .12127

©1996 Kelley Wingate Publications   99   KW 1304

Worksheet 4 (page 100)

Name _____ Skill: Review

Skills Evaluation

Choose the best answer to these review questions. Circle the correct answer.

1. Divide: 43 ÷ 7
- A. 6
- **B. 6 r 1** (circled)
- C. 6 r 3
- D. 6 r 2

2. Divide: 65 ÷ 8
- A. 8
- **B. 8 r 1** (circled)
- C. 8 r 3
- D. 8 r 5

3. Divide: 303 ÷ 4
- A. 76
- **B. 75 r 3** (circled)
- C. 75 r 2
- D. 75

4. Divide: 247 ÷ 5
- A. 9 r 2
- B. 39 r 2
- C. 49 r 1
- **D. 49 r 2** (circled)

5. Find the equal measurement: 5,000 mL =
- A. 500 L
- B. 50 L
- **C. 5 L** (circled)
- D. 0.5 L

6. Find the equal measurement: 6 c. =
- A. 1 qt.
- B. 2 pt.
- **C. 3 pt.** (circled)
- D. 1 gal.

7. Find the equal measurement: 48 hr. =
- A. 1 day
- B. 96 mins.
- **C. 2 days** (circled)
- D. 1 wk.

8. Find the equal measurement: 60 cm. =
- A. 6 mm.
- B. 60 mm.
- **C. 600 mm.** (circled)
- D. 6,000 mm.

9. Find the equal measurement: 5,000 dkm =
- A. 500 km
- **B. 500,000 dm** (circled)
- C. 50 m
- D. 5 cm

10. Find the equal measurement: 240 cg =
- A. 24 mm
- **B. 24,00 g** (circled)
- C. 0.24 dkg
- D. 2,400 dg

©1996 Kelley Wingate Publications   100   KW 1304

1 x 1	1 x 2	1 x 3	1 x 4
1 x 5	1 x 6	1 x 7	1 x 8
1 x 9	10 x 1	12 x 1	2 x 2
2 x 3	2 x 4	2 x 5	2 x 6

4	3	2	1
8	7	6	5
4	12	10	9
12	10	8	6

2 x 7	2 x 8	2 x 9	10 x 2
11 x 2	12 x 2	3 x 3	3 x 4
3 x 5	3 x 6	3 x 7	3 x 8
3 x 9	10 x 3	11 x 3	12 x 3

20 18 16 14

12 9 24 22

24 21 18 15

36 33 30 27

$3\overline{)12}$ $2\overline{)12}$ $5\overline{)10}$ $2\overline{)10}$

$9\overline{)9}$ $3\overline{)9}$ $4\overline{)8}$ $2\overline{)8}$

$6\overline{)6}$ $3\overline{)6}$ $2\overline{)6}$ $5\overline{)5}$

$4\overline{)4}$ $2\overline{)4}$ $3\overline{)3}$ $2\overline{)2}$

5	2	6	4
4	2	3	1
1	3	2	1
1	1	2	1

10 x 8	11 x 8	12 x 8	9 x 9
10 x 9	11 x 9	12 x 9	10 x 10
11 x 10	12 x 10	11 x 11	12 x 11
12 x 12	12⟌12	6⟌12	4⟌12

81 96 88 80

100 108 99 90

132 121 120 110

3 2 1 144

12 x 5	6 x 6	6 x 7	6 x 8
6 x 9	10 x 6	11 x 6	12 x 6
7 x 7	7 x 8	7 x 9	10 x 7
11 x 7	12 x 7	8 x 8	8 x 9

48 42 36 60

72 66 60 54

70 63 56 49

72 64 84 77

4 x 4	4 x 5	4 x 6	4 x 7
4 x 8	4 x 9	10 x 4	11 x 4
12 x 4	5 x 5	5 x 6	5 x 7
5 x 8	5 x 9	10 x 5	11 x 5

28 24 20 16

44 40 36 32

35 30 25 48

55 50 45 40